CAMBRIDGE LIBRARY COLLECTION

Books of enduring scholarly value

Physical Sciences

From ancient times, humans have tried to understand the workings of the world around them. The roots of modern physical science go back to the very earliest mechanical devices such as levers and rollers, the mixing of paints and dyes, and the importance of the heavenly bodies in early religious observance and navigation. The physical sciences as we know them today began to emerge as independent academic subjects during the early modern period, in the work of Newton and other 'natural philosophers', and numerous sub-disciplines developed during the centuries that followed. This part of the Cambridge Library Collection is devoted to landmark publications in this area which will be of interest to historians of science concerned with individual scientists, particular discoveries, and advances in scientific method, or with the establishment and development of scientific institutions around the world.

How to Work with the Spectroscope

John Browning (1830–1925) was the leading British manufacturer of precision scientific instruments, including spectroscopes, telescopes, microscopes, and opthalmoscopes. In *How to Work with the Spectroscope* (1878), he provides a complete overview of the field in which he was the undisputed expert, describing in detail the care and use of instruments ranging from the universal spectroscope to the star spectroscope to the induction coil. This volume also includes Browning's *A Plea for Reflectors* (1867), in which he provides an introduction to the silvered-glass reflecting telescope. Numerous illustrations of the various instruments and a complete price list of Browning's lenses and other apparatuses provide important insight into his business practices and range of expertise. Designed for the lay enthusiast no less than the dedicated scientist, these volumes are also valuable witnesses to the growth of popular science in nineteenth- and early twentieth-century Britain.

Cambridge University Press has long been a pioneer in the reissuing of out-of-print titles from its own backlist, producing digital reprints of books that are still sought after by scholars and students but could not be reprinted economically using traditional technology. The Cambridge Library Collection extends this activity to a wider range of books which are still of importance to researchers and professionals, either for the source material they contain, or as landmarks in the history of their academic discipline.

Drawing from the world-renowned collections in the Cambridge University Library, and guided by the advice of experts in each subject area, Cambridge University Press is using state-of-the-art scanning machines in its own Printing House to capture the content of each book selected for inclusion. The files are processed to give a consistently clear, crisp image, and the books finished to the high quality standard for which the Press is recognised around the world. The latest print-on-demand technology ensures that the books will remain available indefinitely, and that orders for single or multiple copies can quickly be supplied.

The Cambridge Library Collection will bring back to life books of enduring scholarly value (including out-of-copyright works originally issued by other publishers) across a wide range of disciplines in the humanities and social sciences and in science and technology.

How to Work with the Spectroscope

A Manual of Practical Manipulation with Spectroscopes of All Kinds

JOHN BROWNING

CAMBRIDGE UNIVERSITY PRESS

Cambridge, New York, Melbourne, Madrid, Cape Town, Singapore,
São Paolo, Delhi, Dubai, Tokyo

Published in the United States of America by Cambridge University Press, New York

www.cambridge.org
Information on this title: www.cambridge.org/9781108014182

© in this compilation Cambridge University Press 2010

This edition first published 1878
This digitally printed version 2010

ISBN 978-1-108-01418-2 Paperback

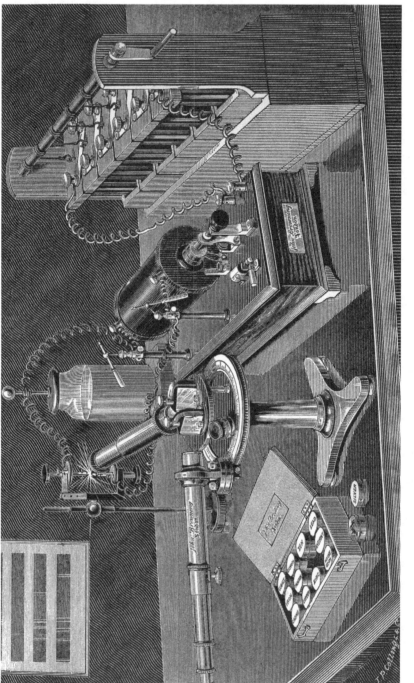

SPECTRUM APPARATUS IN ACTION, SHOWING THE SPECTRA OF THE METALS.

HOW TO WORK WITH THE

SPECTROSCOPE

A MANUAL

OF

PRACTICAL MANIPULATION WITH SPECTROSCOPES OF ALL KINDS,

INCLUDING

DIRECT-VISION	SPECTROSCOPES
CHEMICAL	SPECTROSCOPES
SOLAR	SPECTROSCOPES
STAR	SPECTROSCOPES
AUTOMATIC	SPECTROSCOPES
MICRO	SPECTROSCOPES
SCREEN	SPECTROSCOPES
BESSEMER	SPECTROSCOPES

And Accessory Apparatus.

WITH ABOVE THIRTY ENGRAVINGS AND DIAGRAMS.

BY

JOHN BROWNING, F.R.A.S., F.R.M.S., M.R.I., ETC.

JOHN BROWNING,

63, STRAND, LONDON, W.C.

PRICE ONE SHILLING AND SIXPENCE.

PREFACE.

FOR many years I have had almost daily to reply to inquiries respecting the best method of manipulating with various kinds of Spectroscopes.

It had been suggested to me long since, by our highest authority on the subject, that I should write a *small* work giving the required information. For reasons which will be understood by my friends, I had a great objection to issue such a work, and trusted it would be done by some of the leading scientific men who have worked in this direction. But recently the inquiries for information have been more and more numerous, and I am compelled, both to save my own time and to assist my numerous correspondents, to attempt to supply what appears to be an evident want.

Mr. Proctor's admirable little Manual on the Work of the Spectroscope will be found invaluable by those who do not wish to incur the expense of Roscoe's or Schellen's works on Spectrum Analysis. But Mr. Proctor states that it did not enter into his plans to give detailed instructions for the use of the various kinds of Spectroscopic Apparatus.

I have, therefore, endeavoured to supply such information in the following Pamphlet. The fact that a list of prices is appended of the various Instruments which are described, will not, I hope, be considered to detract from its value. My extensive correspondence leads me to conclude that such information is exceedingly welcome to all those who think of making any experiments for themselves. Once provided with Apparatus, the experimentalist should obtain a copy of Mr. Proctor's book, before alluded to, or of Roscoe's or Schellen's larger works on the subject. Mr. Lockyer's small work on the Spectroscope contains detailed information on the method of working with the Induction Coil, and observing solar prominences.

JOHN BROWNING.

SPECTROSCOPES AND SPECTRUM APPARATUS.

HOW TO WORK WITH A SPECTROSCOPE.

First see that the edges of the slit are free from dust. Direct the slit of the instrument towards the sun, the sky, or some bright light. If there are any horizontal lines visible—*that is, lines running parallel with the spectrum*—they are almost sure to be produced by dust. To remove this dust, open the slit as widely as possible, and wipe the edges of the slit with a small wedge of dry wood. Then close the slit completely ; re-open it, and the lines will probably have disappeared ; if not, repeat the operation. Note that a camel-hair pencil, a leather, cloth, or paper will be sure to *make the slit worse.*

Place the slit as close to the source of light as you can without injuring it.

THE MINIATURE SPECTROSCOPE.

Fig. 1.

Dimensions, $\frac{7}{10}$ diameter, 3 inches long.

This instrument will show many of Fraunhofer's lines, the bright lines of the metals and gases, and the absorption bands in coloured gases, crystals, or liquids.

The Miniature Spectroscope consists of a compound direct-vision prism, which is placed in a sliding drawer. At the end of this drawer there is a lens ; in the best instruments this is an achromatic combination. At the opposite end to the eye-piece there is a slit, which, in the instruments of a superior class, is adjustable by a screw motion. The two jaws move equably from the centre, on rotating the milled ring between the thumb and finger. The jaws of this slit will require to be almost closed to view the Fraunhofer lines, but may with advantage be opened wider when viewing the lines of chemical spectra, or absorption bands in coloured liquids.

HOW TO USE A DIRECT-VISION SPECTROSCOPE.

Fig. 3.

From their simplicity of construction, these instruments are the easiest to use, and are therefore the best adapted for beginners. In using them, it is only necessary to direct the slit to the source of light, which should not be a bright coal-gas or lamp flame, as they will not give any lines. A tallow candle, with a long snuff, will give the yellow or orange sodium line. Coal-gas, burnt in a Bunsen's burner, will give carbon lines. A small quantity of a salt of an alkali or alkaline earth fused on a wire and held in the flame of the Bunsen's burner, will give bright lines. (See " How to Obtain the Bright Lines given by any Substance.") Having by either of these methods obtained some bright lines, focus them carefully by moving the sliding drawer-tube of the telescope, and then close the jaws until the lines appear fine without becoming indistinct.

Any form of stand, with a ring, tube, or a clip to hold the body of the Spectroscope, provided with horizontal, or, better still, with horizontal and vertical motions, greatly facilitates the use of the instrument.

These instructions will apply to the Miniature Spectroscope.

BROWNING'S NEW MINIATURE SPECTROSCOPE, WITH MICROMETER.

Fig. 2.

This portable and complete instrument may be used for showing any of the leading experiments in spectrum analysis ; the Fraunhofer lines ; the lines in the spectra of the metals, and the alkaline earths and alkalies ; the spectra of gases ; and absorption bands.

Applied to a telescope, it may be used for viewing the lines of the solar prominences. It can also be used as a Micro-Spectroscope. The position of the bands in any spectrum may be seen at a glance, as a photographed Micrometer scale is reflected, by means of a magnifying prism, into the field of view, so that it appears parallel with the spectrum. Each tenth line on the scale has a figure above it. This instrument is very convenient for taking the position of a line rapidly.

THE STUDENT'S SPECTROSCOPE.

This instrument has a prism of extremely dense glass of superior workmanship. The circle is divided, and reads with a vernier, thus dispensing with an illuminated scale ; this arrangement possesses the very great advantage of giving angular measures in place of a perfectly arbitrary scale.

The slit is also furnished with a reflecting prism, by means of which two spectra can be shown in the field of view at the same time.

The instrument is so arranged that, with a slight alteration of the adjustments, it can be used for taking the refractive and dispersive powers of solids or liquids. For information on this point see separate heading.

A photographed Micrometer can be applied to this instrument.

Fig. 4.

The instructions for using the Spectroscopes with two to five prisms, which follow, will also apply to this instrument.

THE MODEL SPECTROSCOPE.

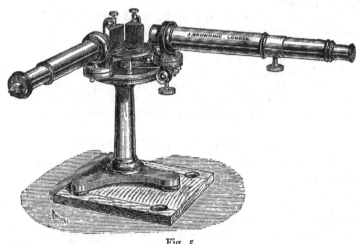

Fig. 5.

This instrument has two dense glass prisms, two eye-pieces, rack motion to telescope, and tangent screw motion to vernier. It will widely separate the D lines.

A photographed Micrometer can be applied to this instrument.

HOW TO USE THE STUDENT'S, THE MODEL, OR THE CHEMICAL SPECTROSCOPE.

Screw the telescope carrying the knife edges at the small end into the upright ring fixed on to the divided circle, and the other telescope into the ring attached to the movable index. Now place any common bright light exactly in front of the knife edges, and while looking through the telescope on the movable index (having first unscrewed the clamping screw under the circle), turn the telescope with the index round the circle until a bright and continuous spectrum is visible.

HOW TO OBTAIN THE BRIGHT LINES IN THE SPECTRUM GIVEN BY ANY SUBSTANCE.

Remove the bright flame from the front of the knife edges and substitute in its place the flame of a common spirit-lamp, or, still better, a gas jet known as a Bunsen's Burner. In the Burner shown in the engraving, there is a ring at the bottom of the tube, with four apertures; by turning these, so that the holes are closed, the Burner gives a white light, well adapted for adjusting the instrument, and showing a continuous bright spectrum in the field of view. When this has been done, turning the ring so as to make the apertures correspond, will admit a quantity of air, and give a dull bluish, very hot flame, well adapted for spectrum analysis. Having obtained this hot flame, take a piece of platinum wire, about the substance of a fine sewing needle; bend the end into a small loop about the eighth of an inch in diameter; fuse a small bead of the substance or salt to be experimented on into the loop of the platinum wire, and attaching it to any sort of light stand or support (as Fig. 28), bring the bead into the front edge of the flame, a little below the level of the knife edges. If the flame be opposite the knife

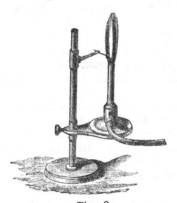

Fig. 28.

Improved Spectroscope Lamp, containing Burner and Clip, on a Single Stand.

edges on looking through the eye-piece of the telescope, the fixed lines due to the substance will be plainly visible. When minute quantities have to be examined, the substance should be dissolved, and a drop of the solution, instead of a solid bead, be used on the platinum wire. The chlorides of the alkaline earths are the best adapted for giving the spectra—as chloride of calcium, chloride of barium, &c.; but carbonate of soda and ferro-cyanide of potassium give better results than the chlorides of these alkalies.

Whatever salts are used, they must be *pure*, as the salts of commerce contain many impurities, and the spectra of these impurities are sufficiently bright to mask the spectra of the substances themselves.

Where coal-gas is not obtainable, a small flame of hydrogen may be used. This should be made in an apparatus in which it is generated only while it is being consumed. The flame of a spirit-lamp may be used, but the results are unsatisfactory. The oxy-hydrogen blowpipe is the best of all the sources of heat known for this purpose; but when the bright lines of the metals are required, the Induction Coil must be used as described under a separate heading.

When it is desired to show the spectrum of a salt for any length of time, small pieces of pumice-stone may be soaked in a saturated solution of the salt, and then attached to a fine platinum wire, and held in the flame as described for a bead.

A frequent source of failure is using a platinum wire too thick for the purpose. The wire should be only just thick enough to support the substance without a tendency to vibration.

The delicacy of this method of analysis is very great. Swan found in 1857 (Ed. Phil. Trans., vol. xxi., p. 411) that the lines of sodium are visible when a quantity of solution is employed which does not contain more than $\frac{1}{2500000}$ of a grain of sodium.

To view Fraunhofer's lines in the solar spectrum, it is only necessary to turn the knife edges towards a white cloud, and make the slit formed by the knife edges very narrow by turning the screw at the side of them. In every instance the focus of the telescope must be adjusted in the ordinary way, by sliding the draw-tube until it suits the observer's sight, and distinct vision is obtained.

It should be noted that lines at various parts of the spectrum require a different adjustment in focussing the telescope.

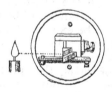

The small prism, turning on a joint in front of the knife edges, is for the purpose of showing two spectra in the field of view at the same time. To do this, it must be brought close to the front of the knife edges. Then one flame must be placed in the position in which the flame of the candle is shown in the small diagram figure, and the other directly in front of the slit. On looking through the telescope as before described, the spectra due to the two substances will be seen one above the other. A known substance should be burnt in the flame, to give one of these spectra, and a material suspected of containing the substance in the other, to see if the lines in the two spectra coincide. This is termed working by comparison, and is the best method of employing the instrument for analytical purposes.

When the slit is turned towards a bright cloud, and a light is used in the position of the candle flame, the spectrum of any substance may be seen, compared with the solar spectrum. In this manner Kirchoff determined in the solar spectrum the presence of the lines of the greater number of the elements which are believed to exist in the sun. The absorption bands in spectra may be most conveniently examined, and accurately investigated, by means of the SORBY-BROWNING Micro-Spectroscope.

TO TAKE THE REFRACTIVE INDEX OR DISPERSIVE POWER OF ANY SUBSTANCE OR LIQUID WITH A SPECTROSCOPE.

A Model or Student's Spectroscope with one or two prisms should be used for this purpose. First remove the prism or prisms from the Spectroscope; bring the telescope in a direct line with the collimator; place a positive eye-piece in the Spectroscope, having cross wires in the field of view; make these cross wires bisect the slit. The substance of which the refractive index is to be determined should be cut into the form of a prism of 60°. Having noted the reading of the telescope on the arc of the Spectroscope, which for this purpose should be divided to degrees and minutes, place the prism and the substance on the plate of the Spectroscope at the minimum angle of deviation—that is, at the angle at which the ray from the collimator is deflected the least. If the solar spectrum be now observed through the telescope and prism, the position of the principal Fraunhofer lines may be read off on the divided arc by bringing them successively to coincide with the cross wires in the eye-piece. The method of calculating the index of refraction and the dispersive power from these observations, will be found very clearly stated in Ganot's "Physics."

When a liquid has to be employed instead of a solid, a hollow prism must be used for containing it. This prism should have glass sides, which can be easily removed for the purpose of cleansing them after each experiment. When in use, the prism and the sides are placed together in a metal frame; a screw in one side of this frame secures them in position, and confines the liquid.

Glass prisms can be made so accurately that the sides, when placed against the prism, will be kept in position by atmospheric pressure; but they are seldom employed on account of their great cost.

BROWNING'S UNIVERSAL AUTOMATIC SPECTROSCOPE.

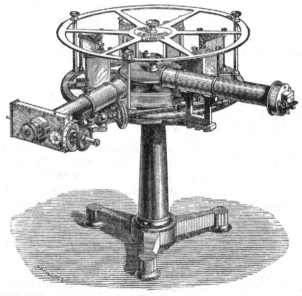

Fig. 7.

In this instrument the prisms are automatically adjusted to the minimum angle of deviation for the particular ray under examination; the position of the lines in the spectra is changed by means of a screw; the revolution of this screw adjusts the prisms automatically for the ray under measurement. The instrument has six prisms, and by means of the reversion of the ray a dispersive power of twelve prisms is obtained. By changing the position of one of the prisms, any dispersive power, from two to twelve prisms, can be used at pleasure, without deranging any of the adjustments of the instrument. This Spectroscope is therefore applicable to every class of spectrum work either in the laboratory or observatory.

Fig. 8.

Diagram showing the Automatic Action by which the Prisms are automatically adjusted to the minimum angle of deviation for the particular Ray under examination.

In an ordinary Spectroscope the prisms are usually adjusted to the minimum angle of deviation for the most luminous rays in the spectrum—by preference, I adjust them myself for the ray E in the solar spectrum. This being done, the prisms are screwed, or otherwise firmly clamped, to the main plate of the Spectroscope. Thus adjusted, they are liable to two sources of error, one of which places the observer at a serious disadvantage. First, only the particular ray for which the prisms have been adjusted is seen under the most favourable circumstances, for only this ray passes, as all should do, through the train of prisms parallel to the base of each prism. Of more importance than this, however, is the fact that the last prism of the train being fixed while the telescope through which the spectrum is viewed is movable around an arc, it is only when the central portion of the spectrum is being examined that the whole field of the object-glass is filled.

Bunsen and Kirchoff, when making their celebrated map

of the solar spectrum, adjusted the prisms they used (four in number) for each of the principal Fraunhofer lines ; but the trouble of doing this is so great that few observers have ever seen the extreme portions of the solar spectrum under favourable circumstances.

The diagram (Fig. 8) shows the method in which the change in the adjustment of the prisms to the minimum angle of deviation for each particular ray can be made automatically. In this diagram there are six prisms. All these prisms, with the exception of the first, are unattached to the plate on which they stand. The first prism is attached by one corner to the plate by a pivot on which it turns. The triangular plates on which the prisms rest are hinged together at the angles corresponding to those at the bases of the prisms. To each of these bases is attached a bar, perpendicular to the base of the prism. As all these bars are slotted and run on a common centre, the prisms are brought into a circle. This central pivot is attached to a dovetailed slide about two inches in length, placed on the under side of the main plate of the Spectroscope, which is slotted to allow it to pass through. On moving the central pivot, the whole of the prisms are moved, each to a different amount in proportion to its distance in the train from the first or fixed prism, on which the light from the slit falls after passing through the collimator, C. Thus, supposing the first prism of the train of C, represented in the diagram, to be stationary, and the second prism to have been moved through 1° by this arrangement, then the third prism will have been moved through 2°, the fourth through 3°, the fifth through 4°, and the sixth through 5°.

A screw gives motion to a lever which is attached to the last prism of the train. By turning this screw until any particular portion of the spectrum appears in the field of view, the rays which issue from the centre of the last prism are made to fall perpendicularly upon the centre of the object-glass of the telescope, T, and thus the ray of light under examination travels parallel to the bases of the several prisms, and ultimately along the optical axis of the telescope itself, and thereby the whole field of the object-glass is filled with light.

Thus the apparatus is so arranged that, on turning the screw so as to make a line in the spectrum coincide with the cross-wires in the eye-piece of the telescope, the lever L, attached to the prisms, sets the whole of the prisms in motion, and adjusts them to the minimum angle of deviation for that portion of the spectrum.

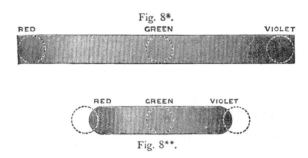

Fig. 8*.

RED GREEN VIOLET

RED GREEN VIOLET

Fig. 8**.

Diagrams 8* and 8** represent the appearances presented looking through the telescope from which the glasses have been removed. In diagram 8* it will be seen that the whole circle of the object-glass is filled with light, as I have just described is the case with the new arrangement; while diagram 8** shows the effect of moving the telescope through the angle in front of the fixed prism.

Here, in the red and violet, *where the light in the spectrum is faintest*, only about one quarter of the field of view is illuminated.

HOW TO USE AN AUTOMATIC SPECTROSCOPE.

Although this Spectroscope is the most powerful, and in appearance the most complex made, it is very easy to use.

To vary the dispersive power of the instrument, let us suppose that the movable right-angled reflecting prism, which reverses the ray, is at the end of the train—that is, the farthest from the collimator. Remove any prism of the battery from its place in the train by sliding it with its holder out of the groove. Take out the reflecting prism in the same way, and slide it into the groove previously occupied by the reflecting prism. In this way the dispersive power of one prism, or of the whole train of prisms, may be employed at pleasure.

The subsequent manipulation will be the same as described under the heading, "How to Use a Chemical Spectroscope," with the exception of mapping the spectra. The method of doing this I have described under the heading, "How to Map a Spectrum with the Automatic Spectroscope."

BROWNING'S AUTOMATIC SOLAR SPECTROSCOPE.

Fig. 9.

Dr. Henry Draper's important discovery of the presence of oxygen in the sun, described in *Nature*, No. 409, August 30, 1877, will direct renewed attention to the solar spectrum.

The Automatic Solar Spectroscope, figured above, will show the solar spectrum with exquisite definition, and if attached to

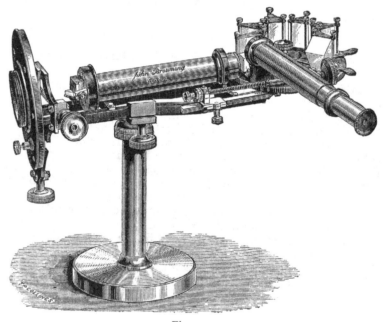

Fig. 10.

the eye-piece of a telescope of three inches or more in diameter, it will show the forms of the solar prominences.

As this Spectroscope can be used with any dispersive power from two to ten prisms, it can be arranged for observing the spectra of the stars and nebulæ. Without a telescope it can be employed for any kind of work in spectrum analysis.

By means of the reversion of the ray this Spectroscope gives a dispersive power equal to ten prisms, and this dispersive power may be changed at pleasure by the observer. The instrument is very light, and can be adapted to a telescope as small as three inches in aperture. It is provided with a movement of rotation for searching for solar prominences.

The Solar Automatic Spectroscope (Fig. 10) is well adapted for use with any telescope, either a reflector or refractor, from six to twelve inches in aperture ; and it can be used on a table stand, as shown in the engraving, for viewing the spectra of metals, salts, or gases.

HOW TO USE A SOLAR SPECTROSCOPE.

This instrument must be used with an astronomical telescope. No eye-pieces must be used with the telescope. The first point to attend to in using a Solar Spectroscope is to place the slit or knife edges of the instrument exactly in the focus of the object-glass or speculum of the telescope.

This must be done by projecting the image of the sun on a card without any eye-piece in the telescope. Note the distance from the end of the telescope, not the end of the sliding drawer, at which a sharp image is produced. Then attach the Spectroscope to the telescope, and move the sliding drawer until the slit is at the required distance from the end of the telescope. The best Solar Spectroscopes are so contrived that the face of the slit can be seen when they are in the telescope. In this case a card may be placed on the face of the slit, and the drawer tube of the telescope moved until the sun's image is seen sharply in focus, when the card may be removed.

Now close the jaws of the slit, and move the telescope, so that the sun's limb or *edge* of the disc falls across the slit at right angles to the direction of the slit. Turn the Spectroscope round, so that every point of the sun's disc passes over the slit, while looking through the instrument. If at the time of making the observation there are any prominences on the sun, a faint red line, and possibly a blue and a green line, will also be visible on looking into the Spectroscope. These lines will appear brighter than the other parts of the solar spectrum.

B

Having obtained the red line, bring it to the centre of the field of view, and carefully open the jaws of the slit while looking at the red line.

If the prominence should be of a well-marked character, its form will then be seen. The problem, when using the Spectroscope, is to get a small bright image into a very narrow slit. In consequence of their large aperture and short focal length, the new Silvered Glass Reflecting Telescopes are admirably adapted for working with the Spectroscope. For information respecting these instruments, see my pamphlet, "A Plea for Reflectors."

THE STAR SPECTROSCOPE.

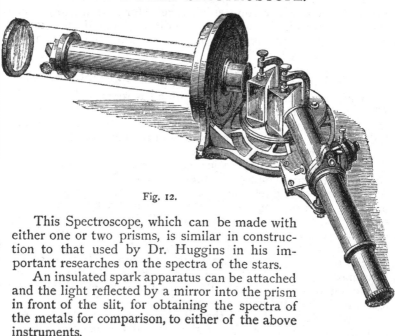

Fig. 12.

This Spectroscope, which can be made with either one or two prisms, is similar in construction to that used by Dr. Huggins in his important researches on the spectra of the stars.

An insulated spark apparatus can be attached and the light reflected by a mirror into the prism in front of the slit, for obtaining the spectra of the metals for comparison, to either of the above instruments.

THE AMATEUR'S STAR SPECTROSCOPE.

This is a Direct-Vision Spectroscope, and is very easy to use with an equatorial provided with a clock; but the new McClean's Star Spectroscope has the great advantage that it can be used on any alt-azimuth stand with the greatest facility.

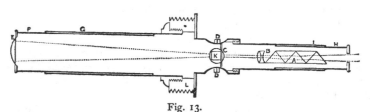

Fig. 13.
Section of Browning's Amateur's Star Spectroscope.

A is a compound direct-vision prism, consisting of five prisms. B is an achromatic lens, which focusses on the slit C, by means of a sliding tube, H ; both the prisms and the lens are fastened in this tube. K is a small right-angled prism, covering half the slit, by the aid of which light may be seen reflected through the circular aperture in front of it. In this manner a comparison may be made with the spectra of metals or gases. The reflecting prism, with the ring to which it is attached, can be instantly removed, and the whole length of the slit used if desired. DD is a ring milled on the edge ; on turning this round, both edges of the slit recede from each other equally, being acted on by two hollow eccentrics. The lines can thus be increased in breadth without their original centres being displaced—a point of importance. E is a cylindrical lens attached to the tube F, which slides in another tube, G. To use the Spectroscope on a telescope, the adapter needs only to have a thread which shall enable it to be screwed into the draw-tube of the telescope, in the place of the ordinary Huyghenian eye-piece. A photographed Micrometer can be added to this instrument.

The draw-tube must then be adjusted so that the slit C comes exactly to the focus of the object-glass. When stars, &c., are about to be observed, this point should be ascertained beforehand, by the aid of an image of the sun, some suitable mark to indicate the focus being made on the draw-tube of the telescope. When this has been once done, the tube can be set by this mark, and the Spectroscope screwed in at any time without any trouble in adjustment.

HOW TO USE A STAR SPECTROSCOPE.

The Spectroscope should occupy the place of an eye-piece in the telescope.

The knife edges or jaws of the slit should be exactly in the focus of the object-glass.

In this position, if the cylindrical lens is removed, the spectrum of a star will be a mere line of light.

The cylindrical lens is for the purpose of widening this line to such an extent that the lines in the spectrum may readily be discerned ; for this purpose the lens must be placed *with its axis at a right angle to the slit*, and the best distance from the slit will be between three and six inches.

The nearer it is brought to the slit the broader will be the spectrum, but it should not be used too close, on account of the diminution of the light.

When it is desired to obtain the spectra of planets, comets, or nebulæ, or indeed any heavenly bodies possessing considerable diameter in the telescope, the cylindrical lens may advantageously be dispensed with.

McCLEAN'S NEW STAR SPECTROSCOPE (PATENT).

Fig. 11.

Many persons well acquainted with the solar spectrum have yet never seen the beautiful and almost infinitely varied spectra of the stars. The Star Spectroscopes in general use are expensive and difficult to manipulate with. This arises from the fact that, in most instruments, the image of a star is required to fall within the jaws of a narrow slit, not more than $\frac{3}{1000}$ inches in width, and an equatorially mounted telescope with clock-work is almost indispensable for using them. Star Spectroscopes of simpler construction, both with and without cylindrical lenses, have been made, but their performance has not been found satisfactory. In the instrument contrived by Mr. McClean, exquisitely fine lines can be seen in the spectra of stars without the use of any slit. The slit being dispensed with, the instrument can be used on any telescope without a clock, or even on any alt-azimuth stand; instead of a slit, a concave cylindrical lens is used, to bring the lines of the spectrum to a focus on the retina.

HOW TO USE McCLEAN'S STAR SPECTROSCOPE.

To use this Star Spectroscope, it is only necessary to bring the star to the centre of the field of view, remove the eye-piece, and insert the Spectroscope instead. The Spectroscope is mounted in a tube, the same size as Browning's Achromatic Eye-pieces; as these slide into an adapter, the change can be effected very quickly, without shaking the instrument. Mr. McClean has adapted a revolving nose-piece to his telescope, one arm of which carries the eye-piece, and another the Spectroscope; and, by the aid of this contrivance, the change from one to the other can be made instantaneously. It is necessary to observe that, this Spectroscope having a negative lens, the lens should be inside the solar focus of the object-glass or mirror employed; roughly, the eye-cap of the Spectroscope should be placed at the solar focus. If the motion of the telescope is not kept up at a proper rate, the light of the spectrum wanes; but being without a slit, the spectrum of a star is never entirely lost, and the observer can, by giving a slight motion to the telescope, either by screws or any other means of adjustment with which the stand may be provided, give the necessary motion to the telescope, to keep the spectrum always as bright as possible.

The Spectroscope is applicable for any telescope of three inches and upwards in diameter.

By throwing the image of an illuminated point into the field of view, any spectra can be seen and used for the purpose of comparison—the required point being illuminated either by a Bunsen's burner, in which salts are being ignited, or by an induction tube, or the electric spark, taken between electrodes of various metals.

Mr. Browning has the exclusive right of making the Spectroscope.

An adjustable slit and convex lens can be used as an addition to McClean's Star Spectroscope, for showing the Fraunhofer lines in the solar spectrum, the bright lines of the metals, alkalies, gases, &c.

HOW TO USE McCLEAN'S UNIVERSAL SPECTROSCOPE.

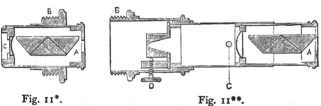

Fig. 11*.   Fig. 11**.

The diagram (Fig. 11*) is a section of McClean's Star Spectroscope as it is used in an astronomical telescope for viewing the spectra of the stars.

When it is desired to use the instrument for chemical purposes, or for showing the Fraunhofer lines, it must be used without a telescope. Take out the small concave cylindrical lens C, which is fitted tightly into the end of the short tube A, containing the prism (Fig. 11*).

Insert the small convex lens, which will be found in the case, in its place.

Remove the adapter B, in which the small tube containing prisms has been used with the telescope.

Now place the tube A, containing the prisms and *convex* lens, in the tube carrying the slit, and proceed to use it as in the instructions for using a Direct-Vision Spectroscope.

The Spectroscope, as now arranged, is shown in the diagram (Fig. 11**). Note that the lines on the tube containing the prisms and the tube carrying the slit must be made to coincide while focussing the lines in the spectrum.

The small milled head D in this diagram serves to regulate the width of the slit.

The second arrangement of the Spectroscope (Fig. 11**) may be used for viewing the chemical spectra, the spectra of the metals, the spectra of gases in induction tubes, or absorption bands in liquids.

Screwed into an astronomical telescope by the coarse screw known as the astronomical thread, B (Fig. 11**), it will show the bright lines of the gaseous nebulæ, the spectra of the planets, or the bright lines of the solar prominences.

THE INDUCTION COIL,

For working Induction Tubes, giving the Spectra of the Gases, or for obtaining the Spectra of the Metals by the aid of the Electric Spark.

Fig. 14.

REQUIRED FOR WORKING INDUCTION TUBES.

Either an Induction Coil, to give a ½-in. spark in dry air, which should be used with 1 quart size Bunsen's cell ;

An Induction Coil, to give a 1-in. spark in dry air, with 2 quart size Bunsen's cells ; or—

An Induction Coil, to give a 1½-in. spark in dry air, with 3 quart size Bunsen's cells.

REQUIRED FOR OBTAINING THE SPECTRA OF THE METALS.

Either an Induction Coil, to give a 2½-in. spark in dry air, with 1 quart size Bunsen's cell ;

An Induction Coil, to give a 3½-in. spark in dry air, with 3 quart size Bunsen's cells ;

An Induction Coil, to give a 4½-in. spark in dry air, with 5 quart size Bunsen's cells ; or—

An Induction Coil, to give a 6-in. spark in dry air, with 6 quart size Bunsen's cells.

Where the trouble of charging Bunsen's cells is objected to, or it is desirable to avoid the nitrous fumes they give off, bichromate batteries can be employed. These batteries are very cleanly, but not nearly so powerful as the Grove's or Bunsen's batteries, so that the Coils will not work with their full power when they are used.

A bichromate battery can be arranged so that, by using a winch, the elements may be removed from the exciting solution at pleasure. These batteries may be used many times without re-charging. One of these batteries is shown in the Frontis-piece.

BROWNING'S SPARK CONDENSER.

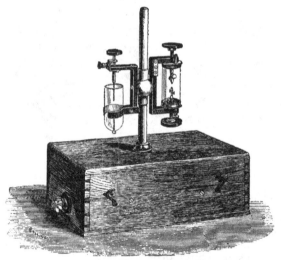

Fig. 15.

This contrivance is designed to replace the Leyden Jars which are generally used with Induction Coils to increase the temperature of the spark when it is required for spectrum analysis. The apparatus consists of an arrangement of ebonite plates coated with tinfoil, and enclosed in a mahogany case. Any amount of surface may be used at pleasure by moving the levers at each end of the case. Unlike the Leyden Jar, the action of the apparatus is not affected by damp. A very convenient arrangement for holding the metals of which the spectra are required, screws on to the lid of the case, and when not in use packs inside the lid.

Becquerel's apparatus for obtaining continuous spectra from solutions of salts of the metals is attached when required. It is shown on the left of the upright rod in Fig. 15.

HOW TO USE BROWNING'S SPARK CONDENSER.

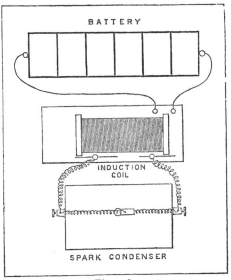

BATTERY

INDUCTION COIL

SPARK CONDENSER

Fig. 15*.

Connect the wires from the battery with the two clamp screws at one end of the Induction Coil. Then carry a fine wire from each of the terminals of the coil (the points from which the sparks are given), one to each clamp screw at the opposite ends of the Spark Condenser. If the commutator of the coil be now turned on, the spark will pass between any pieces of metal placed in the two pairs of tweezers on the insulated ebonite support of the Spark Condenser. This spark will be very different from that of the Induction Coil—being shorter, thicker, and much more brilliant. A spark given through the Condenser, from $\frac{1}{10}$ to $\frac{1}{8}$ of an inch long, is best adapted to give the spectra of the metals in the Spectroscope.

The diagram will perhaps show more clearly than any description how the connections between the battery, spark condenser, and coil are to be arranged. When a Leyden Jar is used, the connections are arranged in a similar manner, but the two wires from the coil must be connected with the inside and outside coating of the Leyden Jar.

HOW TO OBTAIN THE SPECTRA OF THE METALS.

For the purpose of obtaining the spectra of the metals, use an Induction Coil. The coil should give sparks at least two inches long in dry air. Unless you have a Spark Condenser, a Leyden Jar should be introduced in the circuit, as shown in the Frontispiece, for the purpose of increasing the temperature of the spark. Two small pieces of the metal of which the spectrum is required should be placed in forceps attached to the terminals of the Induction Coil. These pieces of metal should be brought within one-eighth of an inch of each other. The spark should pass in a vertical line parallel to and in front of the slit.

The Leyden Jar must be connected with the Induction Coil in the following manner : Attach a wire to the metal rod which supports one pair of forceps on the terminal of the coil, and carry this to the outside covering of the Leyden Jar. A second wire should be attached in a similar manner to the other pair of forceps, and connected with the inside covering of the Leyden Jar. This suffices to bring the Jar into the circuit. When the spark from the metal is obtained the further manipulation will be the same as described, with a bead of salt on a platinum wire, under the heading of the " Chemical Spectroscope."

HOW TO USE BECQUEREL'S APPARATUS.

This Apparatus is shown attached to the Spark Condenser (Fig. 15), on the left of the upright rod. Make a concentrated solution of a salt of the metal. Pour this solution into the glass tube until it covers the platinum wire in the bottom part of the tube by about one-eighth of an inch. Then, by turning the screw with an ebonite head, bring down the upper platinum wire until it is about one-eighth of an inch above the surface of the solution. Having done this, attach one wire from the Induction Coil to the upper platinum wire, and the other to the lower platinum wire. On turning the commutator of the Induction Coil, the spark will pass through the liquid, and on bringing the slit of the Spectroscope close to the side of the tube, the spectrum of the metal which is in solution may be obtained for a considerable time.

LOCKYER'S REVOLVING SPARK APPARATUS.

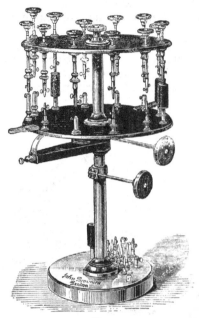

Fig. 29.

This is a contrivance for the purpose of holding a number of pieces of different metals or metallic salts in small carbon crucibles. The metals are held in small pairs of tweezers, each pair being insulated. They are provided with independent adjustments. There is a vertical adjustment by means of a rack and horizontal movement given by an endless screw. The spark only passes through the tweezers in front of the Spectroscope. By means of these adjustments, the various metals can be brought in front of the slit of the Spectroscope, and their spectra obtained with the greatest ease, rapidity, and certainty.

BESSEMER SPECTROSCOPES.

Either of the Direct-Vision Spectroscopes, enumerated on pages 5 and 6, is well adapted for viewing the Bessemer flame, and great numbers are in constant use for this purpose in all parts of Europe, the highest power being best adapted for the purpose ;

but the writer has devised a special instrument (the Direct-Vision Bessemer Spectroscope, with ten prisms), of very great dispersive power, having an eye-piece of large field, which shows the whole of the spectrum, giving admirable definition in all parts of the field.

The instrument below (Fig. 16) is a still more powerful instrument. The telescope has a motion between pivots, near the top of the case. There are cross wires in the field of view to assist the observer in concentrating his own attention, or

Fig. 16.

directing that of others, to any particular line in the spectrum. This instrument is so contrived that the back of the observer is turned to the brilliant flame, which renders vision much easier. A condensing lens, shown in the engraving, which works on a rod in front of the slit, can be fixed so as to produce an image of the flame on the slit ; by moving the instrument about on the hinged joint and swivel, the spectrum of any portion of the flame can be examined at pleasure. With this arrangement the Spectroscope can be used without disadvantage at any distance from the flame. The slit is protected from the action of dust by means of a glass cover, and when the instrument is not in use it can be unscrewed and enclosed in the box.

HOW TO USE A BESSEMER SPECTROSCOPE.

In the Bessemer process several tons of iron are placed in a large vessel, and when in a state of fusion, air is driven through it from apertures in the bottom of the vessel. After about twenty minutes the iron is converted into steel. During the process a flame of almost overpowering brilliancy issues from the mouth of the vessel. The conversion is almost instantaneous, and is known by a change in the colour of the flame and a reduction of its brilliancy. It requires, however, considerable practice to tell by naked eye observations when the process should be stopped ; and intently watching the great glare is most trying to the sight. By the aid of the Spectroscope the completion of the process may be determined without any experience with the utmost ease and certainty.

The spectrum of the Bessemer flame is full of bright lines, a number of green lines being the most brilliant. At the instant of complete conversion these bright green lines suddenly disappear. At this moment the blowing should be stopped. Either of the Direct-Vision Spectroscopes described in this pamphlet may be used for the Bessemer process, but those specially made for the purpose, and termed Bessemer Spectroscopes, will give the most certain results. The engraving (Fig. 16) shows the instrument devised by the writer for Sir John Brown's Steel Works, Sheffield. When using this instrument the observer's back is turned to the bright Bessemer flame. The whole instrument, except the telescope, is enclosed at all times in its mahogany case, and the slit is protected also from the dust caused by the blowing process by a disc of glass.

There is a pointer in the field of view of the eye-piece ; this may be set to any particular line in the spectrum, and the workman may be instructed to stop the process when this disappears.

For full information respecting the spectra given by the Bessemer process, see Roscoe's " Spectrum Analysis."

THE SMALL AUTOMATIC ELECTRIC LAMP.

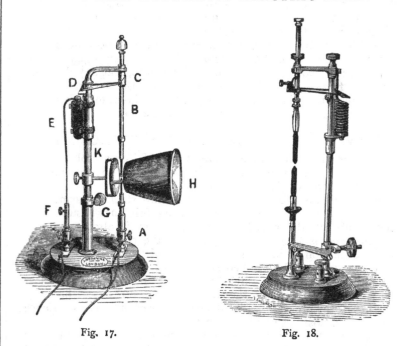

Fig. 17.   Fig. 18.

In the engraving (Fig. 17) the carbon points are carried by the holders, A B, which are provided with rings like a porte-crayon, to clamp the points when in position. C D is a soft iron feeder; the end, C, of this feeder is so arranged that a very slight pressure on the feeder clamps the rod B, and prevents it from descending. E is a rod of soft iron, in the form of a horse-shoe; when the electricity passes through the wire wound upon this horse-shoe, the iron becomes a magnet, and attracts the feeder. F and G are clamping screws, to clamp the sliding rods in any required position. H is a silvered parabolic reflector, for throwing the light of the Lamp to a great distance.

The small Automatic Electric Lamp (Fig. 18) is of similar construction, and can be set in action in the same manner; but it has an additional movement for regulating the height of the carbon poles when burning. This action is influenced by the small milled head on the right-hand pillar, near the base of the stand. Should the poles burn unequally, a few revolutions of this screw will keep them burning always at the same height. This motion is also especially useful when the Lamp is used for showing spectra in screen experiments. In spectrum analysis it should be used in the following manner : The small milled head resting on the top of the arm at right angles to the upright pillar should be turned round, until a spring falls into the oval opening ; it then catches the vertical rod and holds the upper carbon fast. A hollow having been made in the lower carbon, and the metal of which the spectrum is desired having been placed in the small crucible thus formed, the lower carbon is raised by the action of the small screw before referred to, on the right-hand pillar ; when it has been brought into contact with the upper rod, it is carefully separated until the best result is obtained on the screen. More detailed information on this subject will be found under the heading, " How to Show Spectra on a Screen."

The sharpness of the spectrum will be influenced by the width of the slit on the nozzle of the Lantern. The closer the slit the purer will be the colours of the spectrum and the better the definition of the bright lines or absorption bands when liquids are used. Of course the limit is quickly reached at which the slit must be left, or otherwise the results would become unsatisfactory from want of sufficient light.

HOW TO USE THE SMALL AUTOMATIC LAMP.

Release the clamps, F G ; place two pieces of fine hard carbon in the holders ; the carbons should be well pointed ; wipe the rod, B, with a leather, so that it may slide freely ; then adjust the large central rod so that the extreme point of the upper carbon exactly rests upon the lower carbon. Attach the wire from the last plate of zinc in the battery to the lower carbon holder, and the wire from the plate of platinum at the opposite end of the battery to the upper carbon holder. If the light should not burn steadily, alter the position of the magnet by means of a small set screw between the ends ; this screw is not shown in the drawing. The magnet must not be put close to the feeder ; the best distance to place the magnet from the feeder is generally about half an inch, but this will vary with the power of the battery employed.

The Lamp is regulated by means of a small screw, shown in the diagram at C (Fig. 17). This must be done while the battery is attached to the Lamp. By a few trials, turning this screw backwards and forwards, it will be found that the light of the Lamp will become continuous, and it may be left even half an hour at a time, giving a steady light without any attention. Twenty quart cells are quite sufficient to work the Lamp well ; if more are used it becomes too hot. When correctly adjusted there should be no perceptible motion of the feeder.

BROWNING'S LARGE ELECTRIC LAMPS OR REGULATORS.

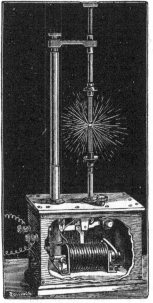

Fig. 19.

In these Regulators both carbons are moved by the electricity of the battery employed (without the aid of clockwork) ; the light remains uniform in height and more steady in action than any of the expensive Regulators previously introduced.

The medium-size Automatic Regulator (Fig. 19).—This Lamp works well with from 20 to 30 pint Grove's cells, or the same number of quart Bunsen's.

The large-size Automatic Regulator.—From 25 to 50 quart Grove's cells, or the same number of two-quart Bunsen's, should be used with this Lamp.

HOW TO USE BROWNING'S LARGE AUTOMATIC LAMP.

As the Electric Lamp is an essential part of Screen Spectrum Apparatus, it is desirable that a brief description of the method of working with the instrument should be given here.

Inside the mahogany case of the Electric Lamp there is a small brass vertical cylinder.

Take care that the piston works freely inside this cylinder, and that the cylinder is full of glycerine.

The object of this arrangement is to prevent the Lamp acting with violent jerks.

Also that the long vertical rod with the milled head at the top moves freely in the tube, and that the iron rods attached to the armature run freely to and fro in the helices of covered copper wire.

When the ends of the carbon rods recede too far from each other, an unsteady light is the result. This can be remedied by moving the small horizontal lever in the front of the large brass plate towards the right. By acting on a strong steel spring, this brings the points of the carbons nearer together.

If the carbon rods do not burn away equally, the light will not be opposite the centre of the parabolic reflector, or lens used for projecting it to a distance. A milled head on the right hand of the base of the Lamp acts on rackwork, which will regulate the height of the lower carbon rod. The upper rod will accommodate itself to the lower rod.

Owing to inequalities in the carbon, the electric light sometimes appears at the side or the back of the carbons. To set this right, move the upper rod by the ebonite head. When a parabolic reflector is used, the light should be brightest on the side towards the upright rod, but it should be brightest on the opposite side to the rod when a lens is used, or when the Lamp is placed in a lantern.

Before making contact with the battery by means of the turning lever of bright copper at the back of the Lamp, press the iron armature firmly forward, so that the iron rods are completely in the hollow coils or helices of covered copper wire.

INSTRUCTIONS FOR CHARGING THE BATTERY.

Fill the porous cells with nitric acid—that is, commercial aqua fortis—and insert the platinum foil or carbon plate. In a strong stoneware vessel, mix one part of oil of vitriol—that is, commercial sulphuric acid—with seven parts of water. Fill the outer cells with this mixture, having first introduced the zinc

c

plates and porous cells. After the porous cells have been placed in the centre of the zinc plates, connect the platinum or carbon plate in each cell with the zinc plate in the next cell by means of the brass clamps; attach one of the clamps with the finger-screw at top to the unconnected platinum or carbon plate at one end of the battery, and the other clamp of the same kind to the unconnected zinc plate at the opposite end of the battery; then connect these ends with the copper wires as before directed. The battery will not attain its full power in less than half an hour after charging. When the battery is done with, the porous cells, zinc, and platinum or carbon plates should be well washed in water. The porous cells should be allowed to remain in fresh water for several hours.

Occasionally, when the zinc plates are taken out of the acid, a little mercury should be well rubbed over them, by means of a piece of rag tied round a small stick. This should be done before they are washed in water.

SETS OF APPARATUS FOR PRODUCING THE ELECTRIC LIGHT.

No. 1.—Small-size Electric Lamp, with Reflector; 20 quart Bunsen's Cells; two Varnished Oak Trays, to hold 10 cells each; Carbon Rods.

No. 2.—Medium-size Automatic Electric Lamp; 30 quart Bunsen's Cells; three Varnished Oak Trays; Carbon Rods.

No. 3. — Large Automatic Electric Lamp; 50 quart Bunsen's Cells; five Varnished Oak Trays; Carbon Rods.

SPECTRUM APPARATUS, FOR SCREEN EXPERIMENTS.

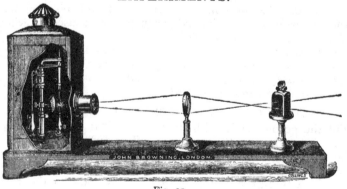

Fig. 20.

The engraving (Fig. 20) represents a new and complete set of apparatus, at a very low price, for projecting the spectra of metals, or the absorption bands of liquids, on a screen. The apparatus comprises an Automatic Electric Lamp and Lantern, with slit, a 20-quart Bunsen's cell, battery, and trays, mounted focussing lens, bisulphide of carbon prism and stand, platform for the whole, and packing case.

An inner case, which fits into the body of the Lantern, contains the Electric Lamp (Fig. 18) in packings, a set of chemicals which give the most brilliant spectra, and a supply of carbon rods and carbon crucibles.

A nozzle with lenses and 3½-inch condensers can be adapted to this Lantern, for showing diagrams or views on screen.

SCREEN SPECTRUM APPARATUS.

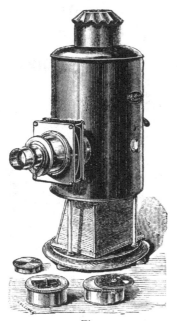

Fig. 21.

Fig. 21 represents a new metal body Electric Lantern for Spectroscopists, with an Electric Regulator specially adapted to the same. There are two nozzles, one for showing diagrams,

and the other for exhibiting spectra on a screen ; this apparatus
is efficient in action, and yet economical in price.

The medium-size Automatic Electric Lamp, which is
adapted to this Lantern, is of the best construction, and works
well with from 20 to 30 pint Grove's cells, or the same number
of quart Bunsen's.

SCREEN SPECTRUM APPARATUS.

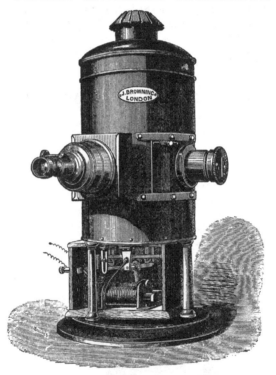

Fig. 22.

Improved Lantern, the body of brass, bronzed, with two
nozzles, especially arranged for exhibiting spectra or diagrams
on the same screen without shifting the Lantern or re-arranging
the apparatus, with 3½-inch condensers (Fig. 22) ; ditto, larger
size, and 4½-inch condensers.

A very complete set of Screen Spectrum Apparatus should
consist of an Improved Electric Camera, with 4½-inch con-

densers (Fig. 22), the body of brass, bronzed, *with two fixed nozzles* specially arranged for exhibiting spectra or diagrams on the same screen, *without shifting the Lantern or re-arranging any part of the apparatus;* large-size Electric Regulator for the above, to work with from 25 to 50 Grove's or Bunsen's cells ; two extra-sized Bisulphide of Carbon Prisms ; Prism Stand and Cover, adjustable for height with clamp motions ; large Condensing Lens on Brass Stand, adjustable for height ; Revolving Diaphragm ; Rotating Carbon Holder; Mahogany Case containing set of Metals and Salts for burning in the Electric Arc, with carbon crucibles, pliers, and 6 feet of carbon rods, &c., for the large Regulator ; Battery of 40 quart Bunsen's cells in four varnished oak trays ; set of Metals and Salts for burning in the Electric Arc, and showing their spectra, carbon crucibles, pliers, and carbon rods, in case.

HOW TO SHOW SPECTRA ON A SCREEN.

The apparatus for screen experiments consists of an Electric Lamp, a Lantern with a slit fixed vertically in place of condensers, a large convex lens mounted on a stand, and either one or two bottle prisms filled with bisulphide of carbon. This apparatus is arranged in the order shown in Fig. 20. The substance of which it is desired to obtain the spectrum should be placed in a hollow cup, bored out of the lower carbon rod in the Electric Lamp or Regulator. The upper rod should be brought into contact with the substance, and then carefully withdrawn until a steady spectrum, consisting principally of bright lines, is seen ; the wider the two carbon rods are separated, so long as sufficient light is left, the better will be the result, as in this manner the brilliant continuous spectrum of the carbon rods, which masks the character of the bright lines, is got rid of. It must be clearly understood that although for all ordinary purposes it is essential that the Electric Lamp should be automatic, yet, for obtaining the spectra of any substance, *the Lamp must be fixed for a certain length of time, for so long as the poles are allowed to approximate to each other, as they would do by the automatic arrangement, so long we shall obtain only the continuous spectrum of the poles, as has been before described.*

The Lamp being arranged within the Lantern, as shown in the Fig., the slit on the nozzle of the Lamp must be placed in a vertical direction ; the fixed lens, adjusted so that the height of the centre of the lens corresponds with the centre of the slit, must then be drawn backwards and forwards along the stand on which the apparatus is fitted, until a sharp image of the slit is produced on a piece of paper, at a distance of from 10 to 20 feet, according to the size of the room in which the experiments are to be conducted ; the distance, however, should be no greater than will produce a spectrum sufficiently large to be seen by the audience. Having obtained the bright image of the slit in the manner just described, either one or two bottle prisms must be brought in front of the lens, just beyond the place at which the rays from the fixed lens cross, after coming nearly to a point. The prism must then be turned round slowly ; as it is turning, a spectrum will be seen to travel along the face of the screen, placed at an angle of nearly 90° to the apparatus, and at about the same distance from the apparatus at which the bright image of the slit was formed. With the small apparatus only one bisulphide of carbon prism can be used ; but with the large Electric Lamp (Fig. 19) and the Lantern (Fig. 22), the light is sufficiently strong to allow of the use of two prisms, and these will give a much longer spectrum. The small carbon crucibles made for the purpose are more efficient than an ordinary carbon rod for obtaining the spectra, as they will hold more of the material to be operated on.

Where time is an object, as in a lecture, a number of these crucibles, fitted in a series of tubes on a revolving holder on the centre round which these revolve, must be placed eccentrically, and the crucibles already charged with the various substances can be brought successively to coincide with the upper pole, and operated on at pleasure. It is difficult to obtain the spectrum of iron in this manner ; a better result can be secured by fixing in the upper carbon holder a piece of iron, and allowing this to come into contact with the carbon rod, taking care that they do not fuze together.

The most interesting experiment which can be made with this apparatus is the reversal of the soda line ; but the experiment is one which requires a little skill in manipulation, yet it is not difficult if the following precautions are attended to : Let the Lamp be carefully closed in with a cover of stiff brown paper over the apertures left for the purpose of ventilation ; set light to a piece of metallic sodium the size of a pea in a small iron spoon, by means of a spirit-lamp ; this must be done inside the Lantern, the spoon being first made nearly red-hot ; the spoon should be attached to a small holder, and left inside the Lamp, with the door shut, until the Lamp becomes filled with the vapour of the burning sodium. One of the carbon crucibles or the lower rod of the Electric Lamp having had a piece of sodium placed on it, the upper end of the Electric Lamp can be brought into contact with it ; the intense light generated will then produce a faint continuous spectrum on the screen, with a very bright sodium line predominant ; after a short time, varying between a few seconds and two or three minutes, if the Lantern has been sufficiently filled with the vapour of the sodium burning in the spoon, these bright lines will become dark ones. Probably in the course of the experiments they will again become bright, and the reversal be repeated two or three times. The explanation of this experiment is, that the vapour of the sodium in the Lantern being cooler than the sodium burning between the poles of the Lamp, the bright light of the sodium is absorbed by the sodium vapour.

Salts of silver, zinc, copper, or small pieces of the metals them-selves, give the most brilliant spectra.

This apparatus may be used without any modification for the purpose of showing the absorption bands in many coloured substances or liquids. In operating on liquids, they should be placed in a hollow wedge-shape cell ; these should be brought in front of the slit, and moved along until such a thickness of the liquid comes in front of the slit as will produce sufficiently strongly-marked absorption bands in the spectrum. During these experiments the Electric Lamp must be kept burning continuously by the automatic action as steadily as possible. The liquids named in the series of specimens on page 43 all give good results. As a few of the most strongly-marked spectra, I will mention blood, sumach, permanganate of potash, chlorophyll, and magenta.

SPECTRUM APPARATUS FOR THE MICROSCOPE.

The writer has worked in conjunction with H. C. Sorby, Esq., F.R.S., in his most recent experiments, having for their object the improvement of this apparatus, and has just per-fected a new Micrometric arrangement, which possesses great advantages. Every line or band in a spectrum, when being measured, is brought to the centre of the field of view ; the jaws of the slit open equally, so that, whatever their width may be, the zero remains unchanged. The Micrometer is self-registering, and whole turns of the Micrometer screw, as well as fractional parts, can be read off at the same place by inspection. The Micro-Spectroscope is applied to the eye-piece of a Microscope instead of an ordinary eye-piece.

It is applicable to opaque objects as well as transparent

without preparation, and by its means two spectra may be compared at the same time with one lamp. It possesses the immense advantage over all other contrivances of the kind, that the spectrum of the smallest object, or a particular portion of any object, may be obtained with the greatest certainty and

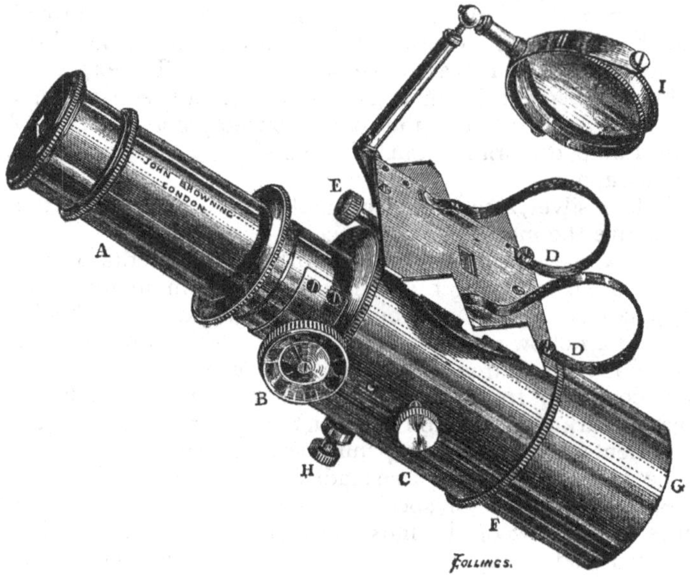

Fig. 23.

facility. This Micro-Spectroscope will indicate plainly the minutest quantity of blood, adulterations in wine, mustard, peppermint, oil, and many other articles of food, as well as the absorption bands in the leaves and juices of plants.

HOW TO USE THE MICRO-SPECTROSCOPE.

As will be seen from Fig. 23, the Micro-Spectroscope is a very compact piece of apparatus, consisting of several parts. The prism is contained in a small tube, A, which can be removed at pleasure. Below the prism is an achromatic eye-piece, having an adjustable slit between the two lenses ; the upper lens being furnished with a screw motion to focus the slit. A side slit, capable of adjustment, admits, when required, a second beam of light from any object whose spectrum it is desired to compare with that of the object placed on the stage of the Microscope. This second beam of light strikes against a very small prism suitably placed inside the apparatus, and is reflected up through the compound prism, forming a spectrum in the same field with that obtained from the object on the stage (Fig. 24, page 46). A (Fig. 23) is a brass tube carrying the compound direct-vision prism, and has a sliding arrangement for roughly focussing. B, a milled head, with screw motion to finally adjust the focus of the achromatic eye-lens.

C, milled head, with screw motion to open or shut the slit *vertically*. Another screw, H, at right angles to C, regulates the slit horizontally. This screw has a larger head, and when once recognized cannot be mistaken for the other.

D D, an apparatus for holding small tube, that the spectrum given by its contents may be compared with that from any other object on the stage.

E, a screw, opening and shutting a slit to admit the quantity of light required to form the second spectrum. Light entering the aperture near E strikes against the right-angled prism which we have mentioned as being placed inside the apparatus, and is reflected up through the slit belonging to the compound prism. If any incandescent object is placed in a suitable position with reference to the aperture, its spectrum will be obtained, and will be seen on looking through it.

F shows the position of the field lens of the eye-piece.

G is a tube made to fit the Microscope to which the instrument is applied. To use this instrument, insert G, like an eye-piece in the Microscope tube. Screw on to the Microscope the object-glass required, and place the object whose spectrum is to be viewed on the stage. Illuminate with stage mirror if transparent, with mirror and lieberkühn and dark well if opaque, or by side reflector, bull's-eye, &c. Remove A, and open the slit by means of the milled head, H, at right angles to D D. When the slit is sufficiently open the rest of the apparatus acts like an ordinary eye-piece, and any object can be focussed in the usual way. Having focussed the object, replace A, and gradually close the slit till a good spectrum is obtained. The spectrum will be much improved by throwing the object a little out of focus.

Every part of the spectrum differs slightly from adjacent parts in refrangibility, and delicate bands or lines can only be brought out by accurately focussing their own parts of the spectrum. This can be done by the milled head, B. Disappointment will occur in any attempt at delicate investigation if this direction is not *carefully attended to.*

When the spectra of very small objects are to be viewed, powers of from $\frac{1}{2}$-inch to $\frac{1}{5}$th may be employed.

Blood, madder, aniline dyes, permanganate of potash solution, are convenient substances to begin experiments with. Solutions that are too strong are apt to give dark clouds instead of delicate absorption bands.

Small cells or tubes should be used to hold fluids for examination.

Objects, such as crystals, should invariably have a small cardboard diaphragm, $\frac{1}{8}$ diameter, placed beneath them; the spectrum is then much better defined. With a slide containing a mass of small crystals, the object need merely be thrown a little out of focus. When observing the spectra of liquids in experiment cells, or through small test-tubes, always slip over the tube containing the $1\frac{1}{2}$ or 2-inch objective a cap with a hole 1-16th of an inch diameter. Slide the tube just sufficiently to bring the small hole a little within the focus of the objective. By this arrangement all extraneous light is prevented from passing up the body of the Microscope, except what passes through the object. Unless this precaution be attended to, a false result is sometimes obtained.

Substances which give bands or lines in the red, are best seen by gaslight, while those which give bands in the blue are brought out far better by daylight. Such a specimen as oxalate of chromium and soda is almost opaque by daylight, showing no bands, though, when examined by a lamp, the spectrum exhibits three beautifully fine lines in the red, two of which are exceedingly delicate. Again, uranic acetate can only be seen to advantage by strong daylight, since the band in the violet would be invisible by lamp-light.

A number of dyes are beautifully shown by being dissolved in gelatine. A plate containing one dozen small strips of gelatine about $\frac{1}{4}$-inch wide and $\frac{3}{4}$-inch long, is exceedingly convenient for the purpose of showing the spectra of these dyes. When the slit of the Spectroscope is placed across the junction of two of the plates, any two spectra can be seen at the same time in the field of view, and thus comparisons may be made between them. If two such plates be superimposed, a great number of spectra, in which the absorption bands of two substances appear at the same time, are shown. These plates are much more easy to manipulate with than tubes.

For information as to mapping spectra with the Micro-Spectroscope, see the instructions under a separate heading.

OBJECTS FOR THE MICRO-SPECTROSCOPE.

Liquids in Glass Tubes Hermetically Sealed, as used by
H. C. SORBY, Esq., F.R.S.

CLASS I.

Specimens for Illustrating the Application of the Micro-Spectroscope to Chemistry.

1. Nitrate Didymium.	7. Cyanide of Cobalt.
2. Uranous Sulphate.	8. Chromic Sulphate.
3. Uranic Acetate.	9. Oxalate of Chromium and Soda.
4. Cobalt in Calcium.	10. Aniline Product, No. 2.
5. Cobalt in Alcohol.	11. Nitrophenic Acid.
6. Aniline Product, No. 1.	12. Uranic Ammonio Carb.

In Morocco Case.

CLASS II.

Specimens for Illustrating the Applications of the Micro-Spectroscope to Vegetable Chemistry.

1. Lobelia Speciosa.	7. Acid Chlorophyll.
2. Red Cineraria.	8. Hypericine, No. 1.
3. Blue Cineraria.	9. ,, ,, 2.
4. Alkanet Root, No. 1.	10. Purpurine from Madder.
5. ,, ,, ,, 2.	11. Camwood.
6. Cudbear.	12. Tradescantia.

In Morocco Case.

CLASS III.

Specimens for Illustrating the Application of the Micro-Spectroscope to Medicine.

1. Cochineal.
2. Acid Cruentine
3. Neutral Cruentine } Blood
4. Deoxidised Hæmaglobin } Compounds.
5. Acid haematin.
6. Deoxidised ,,

In Morocco Case.

CLASS IV.

Specimens to Illustrate the Application of the Micro-Spectroscope to Blowpipe Chemistry and Mineralogy.

BLOWPIPE BEADS AND CRYSTALS.

1. Uranium Oxide.	7. Uranite (Mineral).
2. Chromium Oxide.	8. Acetate of Uranium (Crystals).
3. Copper Oxide.	9. Oxalate of Didymium (Opaque).
4. Cobalt Oxide.	10. Chloride Cobalt (Crystal).
5. Didymium Oxide.	11. Tungsten Sulphide.
6. Permanganate of Potash (Crystals)	12. Molybdenum Sulphide.

Purple Carnet (Ceylon).
In Morocco Case.

CLASS V.

Dyes.

Consisting of a plate containing a set of *twelve* substances in gelatine in a most compact form, and so arranged that *two* spectra can be seen simultaneously by placing the plate in front of the slit of the Spectroscope, giving in all eighteen different comparisons. Also by superposing two such plates at right angles, thirty-six compartments are visible which show the phenomena of mixed spectra.

This arrangement is exceedingly useful for showing absorption spectra at a lecture, since six spectra can be thrown on to the screen at once.

The Plates are fitted in Morocco Cases.

HOW TO MAKE A MAP OF A SPECTRUM WITH A STUDENT'S SPECTROSCOPE.

Place the eye-piece with cross wires in the telescope, with the cross in the direction of an X. Then move the telescope so that the point where the wires bisect comes successively in contact with the various lines, noting the readings of the nonius on the arc. From these readings, by the help of any mechanical scale of equal parts, a map may be easily constructed.

Fig. 26.

HOW TO MAP A SPECTRUM WITH BECKLEY'S SPECTROGRAPH.

Place a sheet of paper on the metal cylinder of the Spectrograph. Note the position of any line in a spectrum, and set the divided edge of the Spectrograph to the corresponding division on the vernier. Draw a line on the paper along the steel straight-edge. Now take the reading of another line, and proceed in a similar manner until the map is completed.

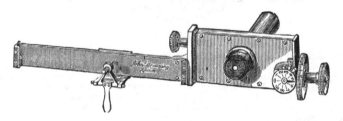

Fig. 27.

HOW TO MAP A SPECTRUM WITH COLONEL CAMPBELL'S SPECTROGRAPH.

Place a strip of smoked glass in the frame attached to the side of the Micrometer (Fig. 27). Place the tube of the Micrometer in the eye-draw of the telescope of a Spectroscope. Bring the cross wires of the Micrometer on to one of the lines in the spectrum by turning the Micrometer screw. Now draw a line on the smoked glass with the small ruling machine on the left hand in the engraving. Repeat this operation for as many lines as it is desired to map, *always turning the screw in one direction.* The smoked glass may be varnished and used as a negative to print copies from by photography, on glass or paper, at pleasure.

HOW TO MAP A SPECTRUM WITH THE AUTOMATIC SPECTROSCOPE.

Place the Filar Micrometer in the telescope. Bring some easily recognized line to correspond with the fixed wire of the Micrometer ; then, by moving the Micrometer-head, make the movable wire coincide with another line in the spectrum. Read the indication on the Micrometer-head, and note it. The small wheel at the side of the Micrometer-head, shows whole revolutions of the screw, the long divisions on the Micrometer-head tenths, and the short divisions hundredths of a revolution. Then take a scale of equal parts, and represent each division on the Micrometer-head by one division on the scale.

It is obvious that the map of a spectrum may be made of any size by varying the scale from which the divisions are taken.

HOW TO MAP A SPECTRUM OF ABSORPTION BANDS WITH A MICRO-SPECTROSCOPE.

Fig. 24 represents the upper part of the Micro-Spectroscope. Attached to the side is a small tube, A A. At the outer part of this tube is a glass plate, blackened, with a fine clear white pointer in the centre of the tube. The lens, C, which is focussed by moving the small studs at M, produces an image of the bright pointer in the field of view by reflection from the surface of the prism nearest the eye. On turning the Micrometer screw, M, the slide which holds the glass plate is made to travel in grooves, and the fine pointer is made to traverse the whole length of the spectrum.

It might at first sight appear as if any ordinary spider's web or parallel wire Micrometer might be used instead of this contrivance ; but on closer attention it will be seen that as the spectrum will not permit of magnification by the use of lenses, the line of such an ordinary Micrometer could not be brought to focus and rendered visible. The bright pointer of the new arrangement possesses this great advantage—that it does not illuminate the whole field of view.

If a dark wire were used and illuminated, the bright diffused light would almost obscure the faint light of the spectra, and entirely prevent the possibility of seeing, let alone measuring, the position of lines or bands in the most refrangible part of the spectrum.

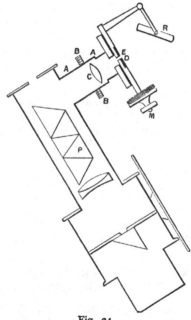

Fig. 24.

To produce good effects with this apparatus, the upper surface of the compound prism, P, must make an angle of exactly 45° with the sides of the tube. Under these circumstances the limits of correction for the path of the rays in their passage through the dispersing prisms are very limited, and must be strictly observed. The usual method of correcting by the outer surface is inadmissible. For the sake of simplicity, some of the work of the lower part of the Micro-Spectroscope is omitted in the engraving. As to the method of using this contrivance : with the apparatus just described, measure the position of the principal Fraunhofer's lines in the solar spectrum. Let this be done *carefully*, in *bright* daylight. A little time given to this measurement will not be thrown away, as it will not require to be done again. Note down the numbers corresponding to the position of the lines, and draw a spectrum from a scale of equal parts. About three inches will be found long enough for this spectrum ; but it may be made as much longer as is thought desirable, as the measurements will not depend in any way on the distance of these lines apart, but only on the micrometric numbers attached to them. Let this scale be done on cardboard, and preserved for reference. Now measure the position of the dark bands in any absorption spectra, taking care for this purpose to use lamplight, as daylight will give, of course, the Fraunhofer lines, which will tend to confuse your spectrum. If the few lines occurring in most absorption spectra be now drawn to the same scale as the solar spectrum, on placing the scales side by side, a glance will show the exact position of the bands in the spectrum relatively to the Fraunhofer lines, which thus treated form a natural and unchangeable scale (see diagram, Fig. 24*). But for purposes of comparison it will be found sufficient to compare the two lists of numbers representing the micrometric measures, simply exchanging copies of the scale of Fraunhofer lines, or the numbers representing them will enable observers at a distance from each other to compare their results, or even to work simultaneously on the same subject.

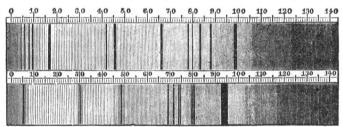

Fig. 24.*

It is a great advantage of this contrivance that it does not monopolize one of the two spectra, as is the case with the use of the quartz scale ; for in describing two spectra only slightly differing from each other, it may be used at once to determine the difference between them. Many substances give two different spectra when examined by transmitted or reflected light, though there is generally a close resemblance between them.

WORKS ON SPECTRUM ANALYSIS.

THE SPECTROSCOPE AND ITS WORK. By RICHARD A. PROCTOR, Author of "Saturn and its System," "The Sun," "The Moon," &c. With Coloured Diagram of the Spectra and many Illustrations. Fcap. 8vo., limp cloth, 1s. (A popular, though by no means superficial, account of the Spectroscope and its marvellous work. For general readers, and as a text book.)

SPECTRUM ANALYSIS in its application to Terrestrial Substances and the Physical Constitution of the Heavenly Bodies, familiarly explained by Dr. H. SCHELLEN, Director der Realschule I. O. Cologne. Translated from Second German Edition by JANE and CAROLINE LASSELL. Edited, with Notes, by WILLIAM HUGGINS, LL.D., D.C.L., F.R.S. In 1 vol., 8vo., with 13 Plates (6 Coloured), including Angström's and Kirchhoff's Maps and 223 Woodcuts. Price 28s. cloth.

PROFESSOR ROSCOE'S LECTURES ON SPECTRUM ANALYSIS (Third Edition), largely illustrated. Six Lectures on Spectrum Analysis and its Applications, delivered before the Society of Apothecaries. Price £1 1s.

ON SPECTRUM ANALYSIS APPLIED TO THE MICROSCOPE. The subject-matter of a Lecture delivered at the South London Microscopical and Natural History Club, by W. T. SUFFOLK, F.R.M.S., with Appendix, &c., six Plates of Absorption Spectra by the Author, and one Chromo-lithograph. Bound in cloth, 3s. 6d.

AN INDEX OF SPECTRA. By W. MARSHALL WATTS, D.Sc. With Preface by Professor ROSCOE, F.R.S. Now ready, demy 8vo., with 8 Lithographic Plates and a Chromo-lithograph. Price 7s. 6d. In this work are collected all the measurements of the spectral lines of the elements, including those by Angström and Thalén, Huggins, Kirchhoff, Plücker, &c., and these are given upon a uniform scale of wavelengths. A drawing of the spectrum of each element is also given.

THE SPECTROSCOPE AND ITS APPLICATIONS. By J. NORMAN LOCKYER, F.R.S. 3s. 6d.

A Coloured Chart of the Spectra of the Alkaline Earths, 36 by 30 inches £0 7 0

A Coloured Chart of the Spectra of the Metals 0 7 0

A Coloured Chart of the Spectra of the Stars and Nebulæ... ... ... 0 7 0

Angström's Normal Spectrum, on which the wave-lengths of the Fraunhofer lines and the lines of the metals are marked in 10,000,000 of a millimetre, in six Plates, with description 0 10 6

NOTE.—The Numbers in this List of Prices correspond with the Numbers appended to the Engravings of the Instruments in the body of the Book.

JANUARY, 1878.

The following Prices are Nett for Cash; half-price allowed for returned Packages, if Carriage Paid.

ORDERS SHOULD BE ACCOMPANIED BY A REMITTANCE.

DIRECT-VISION SPECTROSCOPES.

THE MINIATURE SPECTROSCOPE.

This Instrument will show many of Fraunhofer's lines, the bright lines of the metals and gases, and the absorption bands in coloured gases, crystals, or liquids.

	£	s	d
Miniature Spectroscope, with plain slit	£1	2	0
Miniature Spectroscope, with adjustable slit (Fig. 1)	1	13	0
Morocco Case, extra	0	2	0
Miniature Spectroscope and adjustable slit, with Achromatic Lenses, in Morocco Case	2	6	0
Browning's new Miniature Spectroscope, with Micrometer, price, complete in Case (Fig. 2)	3	10	0

This portable and complete Instrument may be used for showing any of the leading experiments in Spectrum Analysis; the Fraunhofer Lines; the Lines in the Spectra of the Metals, and the Alkaline Earths and Alkalies; the Spectra of Gases; and Absorption Bands.

Applied to a Telescope, it may be used for viewing the Lines of the Solar Prominences. It can also be used as a Micro-Spectroscope. The position of the Bands in any Spectrum may be measured with the attached Micrometer.

	£	s	d
Clip Stand for Miniature Spectroscope	£0	9	6
New Pocket Direct-Vision Spectroscope, in Morocco Case (Fig. 3) ...	4	10	0
New Form Direct-Vision Spectroscope, with five prisms, fitted in Mahogany Case	5	15	0

This Spectroscope is a most powerful and portable Direct-Vision Instrument, easily separating the D lines in the Solar Spectrum.

	£	s	d
Extra power for the above	0	12	6
Large size, higher dispersive power, and extra power eye-piece, complete in Mahogany Case...	6	18	0
Portable Tripod Clip Stand for the above	1	15	0
Direct-Vision Solar Spectroscope, for use with a Telescope for viewing the bright lines or the forms of the Solar Prominences	8	10	0

D

CHEMICAL SPECTROSCOPES.

The Student's Spectroscope, in Stained Cabinet (Fig. 4) £6 10 0

This Instrument has a prism of extremely dense glass of superior workmanship. The circle is divided, and reads with a vernier, thus dispensing with the inconvenience of an illuminated scale ; this arrangement possesses the very great advantage of giving angular measures in place of a perfectly arbitrary scale.

The slit is also furnished with a reflecting prism, by means of which two spectra can be shown in the field of view at the same time.

The Instrument is so arranged that, with a slight alteration of the adjustments, it can be used for taking the refractive and dispersive powers of solids or liquids.

The Model Spectroscope (Fig. 5), with two prisms, in Polished Mahogany
Cabinet £15 0 0

This Instrument has two dense glass prisms, two eye-pieces, rack motion to Tele-scope, and tangent screw motion to vernier.   It will widely separate the D lines.

Photographed Micrometer to either the one or two prism Spectroscope ... £1 15 0

The Model Spectroscope, with four prisms, in superior Cabinet, with
fittings and two eye-pieces 27 10 0

This Instrument is guaranteed to show the Nickel line between the D lines in the solar spectrum.

Browning's Automatic Action, extra £6 10 0

LARGE TABLE SPECTROSCOPES.

The Large Model Spectroscope, for the use of Physicists, made on the
plan of the Gassiot Spectroscope (Fig. 6), in Polished Mahogany
Cabinet £38 10 0

This Instrument has four large very dense glass prisms and Telescopes with object-glasses $1\frac{1}{8}$ in. diameter, and 18 in. focal length, furnished with three eye-pieces.   It will show two or three lines between the D lines in the solar spectrum.   Any smaller number of the prisms can be used when desired.

Dividing ditto on Silver, extra £2 0 0

Browning's Automatic Action to the above 12 10 0

The above Instrument, the circle divided on Silver to 10 seconds, with
five prisms, four eye-pieces, and parallel wire Micrometer, for
measuring the position of lines to $\frac{1}{10000}$ of an inch ; the whole in
Mahogany Case 55 10 0

Browning's Automatic Action to the above 15 0 0

BROWNING'S UNIVERSAL AUTOMATIC SPECTROSCOPE.

In this Instrument the prisms are automatically adjusted to the minimum angle of deviation for the particular ray under examination ; the position of the lines in the spectra is measured by means of a Micrometer ; the revolution of this Micrometer screw adjusts the prisms automatically for the ray under measurement.   The Instrument has six prisms, and by means of the reversion of the ray a dispersive power of 12 prisms is obtained.   By changing the position of one of the prisms, any dispersive power from 2 to 12 prisms can be used at pleasure, without deranging any of the adjustments of the Instrument.   The Instrument is therefore applicable to every class of spectrum work either in the Laboratory or Observatory.

Price of the Universal Automatic Spectroscope, with six prisms, best Filar
Micrometer and Battery of 9 Eye-pieces, in Mahogany Cabinet complete (Fig. 7) £65 0 0

The same Instrument with light framework adapted for application to an Astronomical Telescope £70 0 0

Universal Automatic Spectroscope, dispersion equal to *24 prisms*, 4 Reflecting Prisms, Telescopes 1 foot focus, with McClean's Bright Line Micrometer, light framework ; will give any dispersive power from 2 to 24 prisms 115 0 0

Adapter, with movement of rotation, for attaching the Automatic Spectroscope to a Telescope 6 10 0

Adapter, with two slides, and mechanical motions, to enable the observer to set the Spectroscope at any degree of eccentricity to the Solar disc, so as to sweep either round the Sun's limb to search for prominences, or in the neighbourhood of the Chromosphere 15 0 0

Browning's Universal Automatic Spectroscope, 6 prisms, with the ray inverted, giving a dispersive power of 12 prisms, with the prisms and object-glasses of the Telescopes 1½ in. diameter, and Telescopes 18 in. focal length, Filar Micrometer, and 9 eye-pieces, &c., &c. ... 150 0 0

BROWNING'S AUTOMATIC SOLAR SPECTROSCOPE.

Dr. Henry Draper's important discovery of the presence of oxygen in the Sun, described in *Nature*, No. 409, August 30, 1877, will direct renewed attention to the Solar Spectrum.

The Automatic Solar Spectroscope (Fig. 9) will show the Solar Spectrum with exquisite definition, and if attached to the eye-piece of a Telescope of 3 inches or more in diameter, it will show the form of the Solar Prominences.

As this Spectroscope can be used with any dispersive power from 2 to 10 prisms, it can be arranged for observing the Spectra of the Stars and Nebulæ. Without a Telescope it can be employed for any kind of work in Spectrum Analysis.

By means of the reversion of the ray this Spectroscope gives a dispersive power equal to 10 prisms, and this dispersive power may be changed at pleasure by the observer. This Instrument is very light, and can be adapted to a Telescope as small as 3 inches in aperture. It is provided with a movement of rotation for searching for Solar Prominences.

Complete in Case, with Eye-pieces £28 0 0

Rack Adjustment to the rotary motion, extra 2 10 0

BROWNING'S AUTOMATIC SOLAR SPECTROSCOPE.

By means of the reversion of the ray this Spectroscope gives a dispersive power equal to 10 prisms, and this dispersive power may be changed at pleasure by the observer. It is well adapted for use with any Telescope, either a Reflector or Refractor, from 6 inches to 12 inches in aperture (Fig. 10).

Price complete, with Set of four Eye-pieces... ... ...£42 10 0

Table Stand for using the Automatic Spectroscope above described, without an Astronomical Telescope for viewing the Spectra of Metals, Salts, or Gases 1 15 0

STAR SPECTROSCOPES.

McCLEAN'S NEW STAR SPECTROSCOPE (PATENT).

The Star Spectroscopes in general use are expensive and difficult to manipulate with. This arises from the fact that, in most instruments, the image of a star is required to fall within the jaws of a narrow slit, not more than $\frac{3}{1000}$ inches in width, and an equatorially mounted telescope with clockwork is almost indispensable for using them. Star Spectroscopes of simpler construction, both with and without cylindrical lenses, have been made, but their performance has not been found satisfactory. In the Instrument contrived by Mr. McClean (Fig. 11), exquisitely fine lines can be seen in the spectra of stars without the use of any slit.

Price of the Spectroscope, in Morocco Case...	£2 10 0
Adjustable Slit and Convex Lens, to be used as an addition to McClean's Star Spectroscope, for showing the Fraunhofer lines in the Solar Spectrum, the bright lines of the metals, alkalies, gases, &c., *extra*...	0 18 6
McClean's Spectroscope, for showing both astronomical and chemical spectra, in Case complete...	3 7 6
Rotating Telescope Nozzle, as used by Mr. McClean to carry the New Star Spectroscope, and an astronomical eye-piece, or two eye-pieces of different powers...	1 15 0

This contrivance greatly facilitates the use of McClean's Star Spectroscope, the Spectroscope being carried by one arm, and an eye-piece for observing by the other. Any star seen in the eye-piece can in an instant be examined with the Spectroscope ; or a low power may be used in one arm for finding an object, and a high power for observing it, to be changed rapidly without unscrewing.

STAR SPECTROSCOPES.

Star Spectroscope, with 1 prism, packed in Polished Mahogany Case ...	£8 8 0
Star Spectroscope, with 2 prisms, reflecting prism, to show two spectra at once, and Micrometer Measuring Apparatus for Mapping Spectra, packed in polished Mahogany Case (Fig. 12)...	14 0 0
Insulated Spark Apparatus attached to mirror, for obtaining the spectra of the metals for comparison, adapted to either of the above Instruments	1 1 0
Star Spectroscope of the best construction, with adjustable reflecting prism and mirror, with finest object-glass, Micrometric Apparatus for Measuring the Lines of the Spectrum to $\frac{1}{10000}$ of an inch, extra eye-piece, and ivory tube to reader of vernier, as made for Dr. W. Huggins, F.R.S., packed in polished Mahogany Case, with Insulated Spark Apparatus complete (Fig. 12)	21 0 0
The Amateur's Star Spectroscope, in Mahogany Case (Fig. 13)	4 0 0
Browning's Bright Line Micrometer, for measuring the position of bright lines in the spectra, adapted to the above	2 5 0

INDUCTION COILS,

For working Induction Tubes or obtaining the Spectra of the Metals by the aid of the Electric Spark.

Mr. JOHN BROWNING begs to inform scientific gentlemen that the adoption of an improved method of winding Induction Coils has enabled him to increase their efficiency and reduce their cost. Every Coil is guaranteed to give the length of spark named.

FOR WORKING INDUCTION TUBES.

Induction Coil, to give a ½-in. spark in dry air, with 1 quart size Bunsen's Cell (Fig. 14)	£3 5 0
Induction Coil, to give a 1-in. spark in dry air, with 1 quart size Bunsen's Cell	5 10 0
Induction Coil, to give a 1½-in. spark in dry air, with 2 quart size Bunsen's Cells	6 15 0

FOR SPECTRUM ANALYSIS.

Induction Coil, to give a 2½-in. spark in dry air, with 3 quart size Bunsen's Cells £10 0 0

Induction Coil, to give a 3½-in. spark in dry air, with 3 quart size Bunsen's Cells 12 15 0

Induction Coil, to give a 4½-in. spark in dry air, with 5 quart size Bunsen's Cells 16 0 0

Induction Coil, to give a 6-in. spark in dry air, with 6 quart size Bunsen's Cells 22 0 0

Where the trouble of charging Bunsen's Cells is objected to, or it is desirable to avoid the nitrous fumes they give off, Bichromate Batteries can be supplied. These Batteries are very cleanly, but not nearly so powerful as the Grove's or Bunsen's Batteries, so that the Coils will not work with their full power when they are used.

Bichromate Battery with 4 Cells, arranged so that, by using a winch, the elements may be removed from the exciting solution at pleasure. These Batteries may be used many times without re-charging.

Bichromate Battery, 4 large Cells, lifting elements £4 10 0

Bichromate Battery, 6 large Cells, lifting elements 6 10 0

BROWNING'S SPARK CONDENSER.

This contrivance is designed to replace the Leyden Jars which are generally used with Induction Coils to increase the temperature of the spark when it is required for spectrum analysis. The apparatus consists of an arrangement of ebonite plates coated with tinfoil, and enclosed in a mahogany case. Any amount of surface may be used at pleasure by moving the levers at each end of the case. Unlike the Leyden Jar, the action of the apparatus is not affected by damp. A very convenient arrangement for holding the metals of which the spectra are required, screws on to the lid of the case, and when not in use packs inside the lid (Fig. 15).

Becquerel's Apparatus for obtaining continuous spectra from solutions of salts of the metals is attached when required.

Price of the Spark Condenser for Coils, giving a 2½-inch spark, with levers for changing the number of Plates £2 15 0

Price of the Spark Condenser for Coils up to 5-inch spark, with Levers for changing the number of Plates 3 15 0

Price of the Spark Condenser for Coils giving 6-inch to 8-inch spark ... 9 10 0

Becquerel's Apparatus extra 0 15 0

BESSEMER SPECTROSCOPES.

Either of the Direct-Vision Spectroscopes, enumerated on pages 5 and 6, are well adapted for viewing the Bessemer flame, and great numbers are in constant use for this purpose in all parts of Europe, the highest power being best adapted for the purpose ; but Mr. Browning has devised a special Instrument of very great dispersive power, having an eye-piece of large field, which shows the whole of the spectrum, giving admirable definition in all parts of the field.

Direct-Vision Bessemer Spectroscope, with 10 prisms, complete in Mahogany Case... ... ... £12 10 0

The Bessemer Spectroscope (Fig. 16) is a much more powerful Instrument. The Telescope has a motion between pivots, near the top of the case. There are cross wires in the field of view to assist the observer in concentrating his own attention, or directing that of others, to any particular line in the spectrum. This Instrument is so contrived that the back of the observer is turned to the brilliant flame, which renders vision much easier. A condensing lens, shown in the engraving, which works on a rod in front of the slit, can be fixed so as to produce an image of the flame on the slit ; by moving the Instrument about on the hinged joint and swivel, the spectrum of any portion of the flame can be examined at pleasure. The slit is protected from the action of dust by means of a glass cover, and when the Instrument is not in use it can be unscrewed and enclosed in the box.

Price of the Instrument complete, with Stand £25 0 0

BROWNING'S ELECTRIC REGULATORS.

Small Electric Regulator, with parabolic reflector (Fig. 17). This Regulator will give a powerful and steady light, with from 10 to 20 quart-size Grove's or Bunsen's Cells. Price £2 5 0

Small Electric Regulator (Fig. 18), without reflector, for use in the Lantern, with adjustment for keeping the points of the burning carbons at one height, or separating them to any required distance. This adjustment is indispensable for projecting the spectra of burning metals on a screen. With 20 quart-size Bunsen's Cells, this Regulator will illuminate a 10-feet disc. Price... ... ... 2 15 0

Parabolic Reflector, if required, extra 0 6 6

In these Regulators, described in p. 32, both carbons are moved by the electricity of the battery employed (without the aid of clockwork); the light remains uniform in height and more steady in action than any of the expensive Regulators previously introduced.

Medium-size Automatic Electric Regulator. This Lamp works well with from 20 to 30 pint Grove's Cells, or the same number of quart Bunsen's (as Fig 19). Price £7 10 0

Parabolic Reflector, extra 1 15 0

Large-size Automatic Regulator. From 25 to 50 quart Grove's Cells, or the same number of 2-quart Bunsen's, should be used with this Lamp (Fig. 19). Price 9 9 0

Parabolic Reflector 2 2 0

Carbon Rods, for burning in small Lamp... ... ... per foot 0 1 0

Ditto ditto ... ... per dozen feet 0 10 0

Large Rods, for burning in the large Lamp ... ... per foot 0 1 6

Ditto ditto ... ... per dozen feet 0 15 0

Carbon Cups, for holding metals to obtain their spectra, small, per dozen 0 10 6

Large ditto ditto ditto ditto 0 12 0

GALVANIC BATTERIES.

Grove's Cells	pints (reputed) 11/, quarts (reputed) £0 14 0			
Bunsen's Cells	... pints 5/6, quarts 6/6, 2 quarts 0 9 0			
Varnished Oak Trays	... for 6 cells 6/, for 10 cells 0 10 6			
Bichromate Battery, 4 lifting cells	... large size, £4 10, 6 cells 6 10 0			
Insulated Copper Wire	... ... per yard, 2d. to 0 1 9			
Porous Cells ... per dozen	... ... pints 10/, quarts 0 14 0			
Stoneware Cells ... ,,	... pints 12/, quarts 16/, 2 quarts 1 0 0			
Carbons ... ... ,,	... pints 12/, quarts 16/, 2 quarts 0 18 0			
Zincs ... ... ,,	... pints 12/, quarts 18/, 2 quarts 1 4 0			
Bichromate of Potash	... ...			

BROWNING'S SPECTRUM APPARATUS, FOR SCREEN EXPERIMENTS.

JOHN BROWNING has great pleasure in introducing to the notice of Lecturers and others a New and Complete Set of Apparatus, at a very low price, for projecting the spectra of metals, or the absorption bands of liquids, on a screen. The Apparatus comprises an Automatic Electric Lamp and Lantern, with slit, a Bunsen's battery of 20 quart-size cells, and trays, mounted focussing lens, bisulphide of carbon prism and stand, platform for the whole, and packing case £12 10 0

An inner case can be supplied, which fits into the body of the Lantern, contains the Electric Lamp (Fig. 18) in packings, a set of chemicals which give the most brilliant spectra, and a supply of carbon rods and carbon crucibles ... £4 0 0

Price of the Complete Set of Spectrum Apparatus, packed, £16 10s.

Nozzle with lenses and 3½-inch condensers, for showing diagrams or views on screen, extra £2 15 0

Nozzle with 4½-inch condensers, extra 4 10 0

SCREEN SPECTRUM APPARATUS.

JOHN BROWNING begs to inform Spectroscopists that he has just introduced a new metal body Electric Lantern, with an Electric Regulator specially adapted to the same, for showing diagrams, and exhibiting spectra on a screen; thoroughly efficient in action, and yet economical in price.

The Automatic Regulator is of the best construction, exactly similar to his now well-known large Regulator, but arranged to burn with a smaller number of cells.

Price of the Electric Lantern, for medium-size Lamp, with metal body, japanned bronze green, with two nozzles, interchangeable—one for showing diagrams, with 3½-inch condensers, the other for spectrum analysis (Fig. 21) £7 10 0

MEDIUM-SIZE AUTOMATIC ELECTRIC LAMP
FOR THE ABOVE LANTERN.

This Lamp works well with from 20 to 30 pint Grove's cells, or the same number of quart Bunsen's (Fig. 19) £7 10 0

LARGE ELECTRIC LANTERNS.

Improved Lantern, the body of brass, bronzed, *with two nozzles*, specially arranged for exhibiting spectra or diagrams on the same screen *without shifting the Lantern or re-arranging the apparatus*, with 3½-inch condensers (Fig. 22) £11 10 0

Ditto, ditto, larger size, and 4½-inch condensers 16 10 0

Large Electric Lamp (Fig. 18) for either of the above Lanterns 9 9 0

VERY COMPLETE SET OF
SCREEN SPECTRUM APPARATUS.

Improved Electric Camera, with 4½-inch condensers, the body of brass, bronzed, with two nozzles specially arranged for exhibiting spectra or diagrams on the same screen, without shifting the Lantern or re-arranging the apparatus. Large-size Electric Regulator for the above, to work with from 25 to 50 Grove's or Bunsen's cells. Two extra-sized Bisulphide of Carbon Prisms. Prism Stand and Cover, adjustable for height with clamp motions. Large Condensing Lens on Brass Stand, adjustable for height. **Revolving Diaphragm. Rotating Carbon Holder.** Mahogany Case containing set of metals and salts for burning in the Electric Arc, with carbon crucibles, pliers, and 6 feet of carbon rods, &c., for the large Regulator. Battery of 40 quart Bunsen's cells in four varnished Oak Trays.

Price of the Set of Apparatus, complete £50	0	0
Browning's New Lantern Microscope adapted to the above for showing Microscopic Objects on a Screen...	5	10	0
Set of Metals and Salts for burning in the Electric Arc, and showing their spectra, with carbon crucibles, pliers, and 5 feet of carbon rod, in Mahogany Case	2	10	0

SETS OF APPARATUS FOR PRODUCING THE ELECTRIC LIGHT.

No. 1.

Small-size Electric Lamp, with Reflector (Fig. 17)... ... ... £2	5	0
20 quart Bunsen Cells, at 6/6 each	6	10	0
2 Varnished Oak Trays, at 10/6 each	1	1	0
Carbon Rod ... ... per foot 1/, per dozen feet	0	10	0

No. 2.

Medium-size Automatic Electric Lamp, without Reflector (Fig. 19) ...	7	10	0
Parabolic Reflector, extra	2	2	0
30 quart Bunsen's Cells, at 6/6 each	9	15	0
3 Varnished Oak Trays, at 10/6 each	1	11	6
Carbon Rods ... ... per foot 1/6, per dozen feet	0	15	0

No. 3.

Large Automatic Electric Lamp (Fig. 19)	9	9	0
Parabolic Reflector, extra	2	2	0
50 quart Bunsen's Cells, at 6/6 each	16	5	0
5 Varnished Oak Trays, at 10/6 each	2	12	6
Carbon Rods ... ... per foot 1/6, per dozen feet	0	15	0

SPECTRUM APPARATUS FOR THE MICROSCOPE.

Mr. BROWNING has worked in conjunction with H. C. SORBY, Esq., F.R.S., in his most recent experiments, having for their object the improvement of this apparatus, and has just perfected a new Micrometric arrangement, which possesses great advantages. Every line or band in a spectrum, when being measured, is brought to the centre of the field of view ; the jaws of the slit open equally, so that, whatever their width may be, the zero remains unchanged. The Micrometer is self-registering, and whole turns of the Micrometer screw, as well as fractional parts, can be read off at the same place by inspection. The Micro-Spectroscope is applied to the eye-piece of a Microscope instead of an ordinary eye-piece.

It is applicable to opaque objects as well as transparent without preparation, and by its means two spectra may be compared at the same time with one lamp. It possesses the immense advantage over all other contrivances of the kind, that the spectrum of the smallest object, or a particular portion of any object, may be obtained with the greatest certainty and facility.. This Micro-Spectroscope will indicate plainly the minutest quantity of blood, adulterations in wine, mustard, peppermint, oil, and many other articles of food, as well as the absorption bands in the leaves and juices of plants.

Price of the Micro-Spectroscope, complete with Micrometer £8 5 0
The Sorby-Browning Micro-Spectroscope, with rack-motion to eye-piece (Fig. 24) without Micrometer 6 0 0
Browning's Bright-Line Micrometer, for measuring the position of bright lines in Spectra (Fig. 24) £2 5 0
Case for Spectroscope, with Racks for Cells and Tubes 0 15 0
Sorby's Tubes ... ... per doz. 0 2 6
Sorby's Wedge Cells ... ... per doz. 0 6 0
Specimens in Sealed Tubes for showing the bands, each 0 1 6

For List of Specimens for the Micro-Spectroscope, see page 43.

Price of the Specimens—Class I. and II., £1 1s. ; Class III., 11s. 6d.; Class IV., £1 1s. ; Class V., 6s. 6d.

The Amateur's Micro-Spectroscope, with Achromatic Lens and Reflecting Prism, to show Two Spectra at the same time, for the purpose of comparison 2 15 0
Mahogany Case for the above 0 5 0

TO LECTURERS ON SCIENCE.

JOHN BROWNING begs to announce that he has prepared, with peculiar care, a great number of Diagrams, principally Photographs, to illustrate recent discoveries in Spectrum Analysis and other branches of Observational Astronomy. These slides can be had either plain or exquisitely coloured.

Prices, Plain £0 3 6
Coloured ... ... from 4/6 to 0 10 6
Photographs of Microscope Objects for the Lantern, each... ... ... 0 3 6

For List of Subjects, see end of Catalogue.

DENSE GLASS PRISMS.

Prisms of extra dense flint, of very superior quality, of 45 or 60 degrees, with accurate plane surfaces, ¾ inch £0 15 0
Prisms, 1 in., 20/; 1¼ in., 30/; 1½ in., 60/; 2½ in., 90/; 2¾ in. 6 0 0
Prisms, 3 in., £15 ; 4 in. by 3 in. 30 0 0
Prisms of the above sizes, of the densest flint made, 25 per cent. extra.

BISULPHIDE OF CARBON PRISMS.

Bisulphide of Carbon Prisms, large size £0 15 0
Bisulphide of Carbon Prisms, extra large size 0 18 6
Bisulphide of Carbon Prisms, with Parallel Glass Sides, Browning's improved method of Mounting in Metal Frames ... ... each 1 15 0

Prisms of other Angles to Order.

HOLLOW GLASS PRISMS.

Hollow Glass Prisms, with movable sides in metal frames, for taking the refractive index or dispersive power of a liquid, angles of 60°, 1¼ in. faces, to replace the usual dense glass prism on a Spectroscope at pleasure ... ... each £0 18 6

Hollow Glass Prisms, with perfectly plane and parallel sides, without metal frames, warranted to give the finest definition ...from £3 3s. to 6 6 0

SPECTROGRAPHS.

Beckley's Spectrograph, for Mapping out Spectra (Fig. 26)	£10	0	0
Beckley's Spectrograph, larger	15	0	0
Colonel Campbell's Automatic Micrometric Spectrograph (Fig. 27) ...	16	10	0

SUNDRY SPECTROSCOPIC APPARATUS.

Hollow Cells, with one side formed of a prism, for holding solutions for examining absorption-bands £1 1 0

Large ditto, for projecting Spectra on Screen 1 11 6

Extra power Eye-pieces ... ... 12s. 6d. to 1 0 0

Bunsen's Burners ... ... 3s. 6d. to 0 5 0

Adjusting Clip, on stand, to hold platinum wires 0 3 6

Browning's Improved Spectroscope Lamp, containing burner and clip on a single stand, complete (Fig. 28) 0 12 6

Brass Stand, superior finished ditto 1 15 0

Leyden Jars ... ... from 3/6 to 2 2 0

Insulated Spark Apparatus, on brass stand, with 2 Dischargers for obtaining the spectra of metals and gases 1 18 6

Set of 13 chemically pure Metals, in Mahogany Cabinet, for Spectrum experiments 0 18 6

Vacuum Tubes, prepared for showing the beautiful Spectra of various Gases—Nitrogen, Hydrogen, Oxygen, Carbonic Acid, Ammonia, Sulphuric Acid, Olefiant, Chlorine, Bromine, Iodine, Coal Gas, Æther Vapour, Turpentine Vapour, Petroleum Oil Vapour, and Water Vapour ... ... each, 5/6, 7/6, and 0 8 6

Set of Salts best adapted for showing Chemical Spectra, stoppered bottles, in case 0 7 6

Metallic Thallium and other Chemicals to order.

Platinum Wire, for use with the Spectroscope ... ... per foot 0 1 0

Plucker's Tube Holder, for holding a single Plucker's Tube 1 5 0

Plucker's Tube Holder, for holding 7 tubes 3 3 0

Insulated Spark Apparatus, on brass stand, for obtaining the Spectra of Metals 1 5 0

Lockyer's Insulated Spark Apparatus, with vertical and horizontal rack motions 1 15 0

Lockyer's Insulated Spark Apparatus, with 7 Dischargers for obtaining the spectra of 7 metals without altering the Apparatus (as Fig. 29) .. 4 0 0

Lockyer's Insulated Spark Apparatus, with 14 dischargers (Fig. 29) ... 6 10 0

Reflecting Mirror for following the motion of the sun, so as to investigate or map the Solar Spectrum. This is a rectangular mirror of large size, mounted on an axis, and provided with endless screw motions in altitude and azimuth, and Hook's joints, with long handles, which can be carried to the foot of the Spectroscope when the mirror is placed outside of a window. Price £8 10 0

The following prices are Nett for Cash ; half-price allowed for returned Packages, if Carriage Paid.

ORDERS SHOULD BE ACCOMPANIED BY A REMITTANCE.

———

LIST OF PRICES,

JANUARY, 1878.

———

SILVERED GLASS TELESCOPES AND SPECULA.

SILVERED GLASS SPECULA, UNMOUNTED.

WITHOUT CELLS.

The performance of these Specula will be guaranteed ; they will bear a power of 100 to the inch on suitable objects and under favourable conditions of the atmosphere.

Speculum	4½	inch diameter, about	5 ft. focus	...	...	...	...	£5	0	0
,,	6½	,,	,,	5 ft. or 6 ft. focus	...	...	...	9	7	0
,,	8⅜	,,	,,	5 ,, 8 ,,	...	...	...	17	12	0
,,	9¼	,,	,,	6 ,, 8 ,,	...	...	...	23	2	0
,,	10¼	,,	,,	6 ,, 9 ,,	...	...	...	38	10	0
,,	12¼	,,	,,	6 ,, 10 ,,	...	...	...	55	0	0
,,	13	,,	,,	9 ,, 10 ,,	...	...	...	82	10	0
,,	15	,,	,,	,, 10 ,,	...	...	...	110	0	0
,,	18	,,	,,	,, 15 ,,	...	...	...	150	0	0

PRICES OF SILVERED GLASS SPECULA ASTRONOMICAL TELESCOPES, ON ALT-AZIMUTH STANDS.

3½ inch Speculum, 3 ft. focus, mounted in metal, on metal alt-azimuth stand, with two eye-pieces, 50 to 150 ... ...£10 15 0

4½ inch Speculum, 5 ft. focus, mounted on a stand, which can be changed from alt-azimuth to parallactic, so that the stars can be followed with one motion, with endless driving screw, and Hook's joint and two eye-pieces, 100 to 200 (Fig. 5) in " Plea for Reflectors " 24 4 0

6½ inch Speculum, 6 ft. focus, on alt-azimuth stand, with quick and slow fine screw motions, and three eye-pieces, 100 to 450 (Fig. 6) ... 36 6 0

8½ inch Speculum, 8 ft. focus, mounted as above, with three eye-pieces, 100 to 500 (Fig. 6) 49 5 0

9¼ inch Speculum, 8 ft. focus, as above, with four eye-pieces, 100 to 600 (Fig. 6) 59 10 0

10¼ inch Speculum, 9 ft. focus, ditto (Fig 6) 79 6 0

SILVERED GLASS SPECULA ASTRONOMICAL TELESCOPES,

EQUATORIALLY MOUNTED IN A SUPERIOR MANNER.

4½ inch Speculum, 5 ft. focus, equatorially mounted (angle for latitude to order), with 6 inch hour circle reading to 5 seconds, and declination circle reading to 1 minute, two eye-pieces, 100 and 300 (Fig. 7) ...£49 10 0

6½ inch Speculum, 6 ft. focus, with 12 inch hour circle reading to 5 seconds, and declination circle to 1 minute, three eye-pieces, 100 to 450, rotating hour circle	£88 0 0
8½ inch Speculum, 8 ft. focus, mounted as above	115 10 0
9¼ inch Speculum, 8 ft. focus, with four eye-pieces, 100 to 600	148 10 0
10¼ inch Speculum, 9 ft. focus	181 10 0
12¼ inch Speculum, with extra eye-pieces	242 0 0
13 inch Speculum, with ten eye-pieces, including Achromatics and Kellner	368 10 0
15 inch Speculum, 10 ft. focus, with 16 inch hour circle, reading to five seconds and declination circle to one minute, with three Huyghenian eye-pieces, one Kellner, and six Browning's improved Achromatic eye-pieces, powers ranging from 60 to 600 diameters. Position Micrometer, with two verniers divided on silver, and reading to single minutes, with clock-work driving apparatus complete...	650 0 0
Clock-work Driving Apparatus to 8½, 9¼, or 10¼ inches	38 0 0
Ditto ditto 12¼ or 13 inches	49 0 0

These Instruments can be furnished with Reflecting Prisms of the finest quality, in place of the diagonal mirrors generally used; but Silver Planes are recommended and supplied as giving the finest definition. When planes are chosen, two of the choicest quality will be sent with each Instrument.

SILVERING GLASS SPECULA.

4½ inches	£0 5 0	13 inches	£1 10 0
6½ „	0 10 0	15 „	2 0 0
8½ „	0 12 6	Silvering Plane Mirrors ...	0 2 6
9¼ „	0 15 0	Silvering above 2½ inches in the minor axis	0 3 6
10¼ „	1 0 0		
12¼ „	1 5 0		

All charges incurred for carriage will be extra.

ASTRONOMICAL EYE-PIECE—HUYGHENIAN CONSTRUCTION.

Nos. 1 and 2, magnifying 65 and 85	£0 15 0
„ 3, 4, and 5, „ 125, 200, and 250	1 0 0
„ 6 „ 400	1 5 0
„ 7 „ 600	1 10 0

ACHROMATIC EYE-PIECES.

These Eye-pieces have a rather limited field, but their performance with reflecting Telescopes, particularly on planets, is very superior to Huyghenian.

A Magnifying 86	£1 2 6	E Magnifying 306	£1 15 0
B „ 144	1 10 0	F „ 450	2 0 0
C „ 208	1 15 0	G „ 600	2 5 0
D „ 250	1 15 0	H „ 840	2 10 0

LARGE FIELD EYE-PIECES.

Very low power Comet Eye-piece, magnifying 35	£1 0 0
Kellner Eye-piece, with field of 50 minutes, for clusters or nebulæ, magnifying 60	1 7 6
Ditto, with field of 35 minutes, for clusters, nebulæ, or the moon, magnifying 85	1 7 6
Day-power Eye-pieces, erect .. .. £1 0 0 to	1 10 0

The power of all the Eye-pieces has been calculated on an object-glass, or mirror, of six feet focus.

SOLAR EYE-PIECES.

	£	s.	d.
Single Reflecting Prism, mounted for viewing the Sun or Moon, with two shade heads, in Mahogany Case	£3	10	0
Ditto, with 2 Prisms, arranged for single reflections, for viewing the Sun only	8	0	0
Browning's new Solar Eye-Piece, single	4	0	0
Ditto, with 2 Double Prisms	8	0	0

TRANSIT EYE-PIECES.

	£	s.	d.
Transit Eye-Piece, with fine webs or wires	1	5	0
Ditto, with 7 webs and higher powers ... ...£1 10 0 to	1	15	0

PARALLEL WIRE MICROMETER.

	£	s.	d.
Micrometer for measuring to seconds ... ...£6 to	11	11	0

POSITION MICROMETER.

	£	s.	d.
Parallel Wire Micrometer, with position circle and two verniers, reading to single minutes	11	0	0
Ditto, superior make, divided on silver ... ·	13	15	0
Ditto, divided on Platinum	17	0	0

Extra Eye-Pieces, 15s. each.

	£	s.	d.
Browning's Double Image Micrometer	8	10	0

BARLOW'S LENS.

This is an achromatic combination of a negative focus. On inserting it behind any eye-piece (that is, between the eye-piece and the object-glass), the power of the eye-piece is increased from one-third to one-half; at the same time the introduction of this lens, especially when using Huyghenian eye-pieces, greatly improves the performance of reflecting Telescopes, especially on bright stars.

Price of the best quality, £1 2s. 6d.

	£	s.	d.
SMOKE-COLOURED GLASS WEDGES corrected for refraction for intensifying the marks on the Moon or Planets	£1	2	0
NEUTRAL TINT WEDGES for observing the Sun	1	2	0
DIVIDED LENS DYNAMETER, for measuring accurately the power of eye-pieces, new movement for separating the lenses	4	4	0

ASTROMETER.

	£	s.	d.
Knobel's Astrometer, a simple and efficient Instrument for determining Star magnitudes	12	10	0

PERFECT PLANES, UNMOUNTED.

PERFORMANCE UNDER ANY POWER GUARANTEED.

	£	s.	d.
1 inch in the minor axis of the ellipse	1	0	0
1½ ,, ,, ,, ...	1	10	0
2 ,, ,, ,, ...	2	0	0
2½ ,, ,, ,, ...	2	10	0
3 ,, ,, ,, ...	3	5	0

DIAGONAL PRISMS, Unmounted.

With circular surfaces on the planes at right angles to each other. The diameter of these planes is equal to the minor axis of an ellipse which would be required if a mirror were employed.

Plane Surfaces ¾ inch in diameter			...	...	...	...	...	...	£1	3	0
,,	1	,,	...	...	...	...	...	...	2	4	0
,,	1¼	,,	...	...	...	...	...	...	3	9	0
,,	1½	,,	...	...	...	...	...	...	5	15	0
,,	2	,,	...	...	...	...	...	...	8	0	0
,,	2½	,,	...	...	...	...	...	...	11	11	0

These Prisms are made of a very pure hard white crown glass. They reflect more light, and that freer from colour, than silvered diagonal mirrors. They are not liable to injury from moisture. Prisms possess the foregoing advantages over plane mirrors ; but for dividing the most difficult double stars, a plane mirror is *decidedly the best.*

Prepared Pad for polishing Specula, in bottle, 2/6.

An Illustrated Catalogue of Spectroscopes sent post free for 7 stamps.

,,   ,,   ,,   *Microscopes*   ,,   ,,   7   ,,

ASTRONOMICAL WORKS.

A New Star Atlas for the Observatory—12 Maps with letter-press—by R. A. Proctor, B.A., F.R.A.S.	£1	5	0
Descriptive Astronomy, by F. Chambers, F.R.A.S.	1	8	0
Celestial Objects, by Rev. T. W. Webb, M.A., F.R.A.S. (Third Edition)	0	7	6
Elementary Astronomy, by J. N. Lockyer, F.R.A.S.	0	5	6
Saturn and its System, by R. A. Proctor, B.A., F.R.A.S.	0	14	0
Half-hours with the Stars, by R. A. Proctor, B.A., F.R.A.S.... ...	0	5	0
Half-hours with the Telescope, by R. A. Proctor, B.A., F.R.A.S. ...	0	2	6
Sun Views of the Earth, by R. A. Proctor, B.A., F.R.A.S.	0	5	0
Other Worlds than Ours, by R. A. Proctor, B.A., F.R.A.S. (Second Edition)	0	10	6
Handbook of the Stars, by R. A. Proctor, B.A., F.R.A.S.	0	5	0
The Sun, by R. A. Proctor, B.A., F.R.A.S.	0	14	0

"A Plea for Reflectors ;" *being a Description of the new Astronomical Telescopes, with Silvered Glass Specula, and Instructions for Using and Adjusting them, with many Illustrations, and Coloured Frontispiece of Jupiter* (Sixth Edition), *by* John Browning, F.R.A.S.—*One Shilling, post free.*

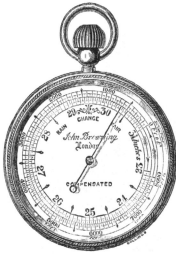

BROWNING'S GRAPHOSCOPES.

Handsome Graphoscope, in Walnut and Maple, with Viewing Lens, 5 inches in diameter, and Stereoscope, complete £3 3 0

(The Stereoscope shown in use.)

Illustrated List of Instruments sent Post Free by

JOHN BROWNING, 63, Strand, London.

ESTABLISHED ONE HUNDRED YEARS.

BROWNING'S MICROSCOPES.

The Model Microscope, with Rack and fine Adjustments to body, with
Axis for inclination, one Eye-piece, Concave Mirror, and 1-inch and
¼-inch Object-glasses £5 10 0

List of Microscopes sent Free.
Illustrated Catalogue of Microscopes for Six Stamps.

JOHN BROWNING, 63, Strand, London.
ESTABLISHED ONE HUNDRED YEARS.

BROWNING'S NEW BINOCULARS.

The "PANERGETIC" Opera, Field, and Race Glass,

For general use, brilliant light, extensive field of view, and sharp definition.

NOTICES OF THE PRESS.

"Brings out figures with marvellous distinctness, and has a very large field of view, and so many advantages over the other Binoculars that we have seen, that we confidently award very high praise indeed."—*Popular Science Review.*

"Exhibits objects with remarkable brightness and sharpness."—*The Observer.*

"A wide extension of the field of view is attained, while even in misty weather objects are exhibited with wonderful clearness."—*Naval and Military Gazette.*

"A very extensive field of view is obtained, and objects in the distance are shown with great distinctness."—*English Churchman.*

Price £4 10s.

The "EURYSCOPIC" Opera, £2 2s.

For the Theatre, has the largest field of view, giving delightfully easy vision.

ILLUSTRATED LIST FREE BY POST.

ACHROMATIC OPERA GLASS in Case, from 10s. 6d.

JOHN BROWNING,

Optical and Physical Instrument Maker to H.M. Government, the Royal Society, the Royal Observatories of Greenwich and Edinburgh, and the Observatories of Kew, Cambridge, Durham, Utrecht, Melbourne, &c.

63, STRAND, W.C. (Factory, Southampton Street, London, W.C.)

ESTABLISHED ONE HUNDRED YEARS.

INDEX.

INDEX (continued).

WARREN DELARUE.

"SATURN".
Seen in a 13 inch metallic reflector.

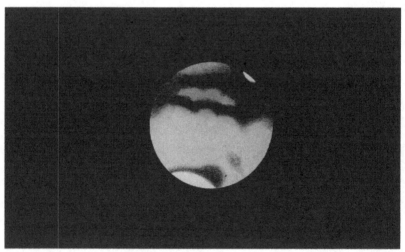

J BROWNING

Francis & Jackson lith London.

"MARS".
Seen in an 8½ inch silvered glass reflector.

A PLEA FOR

REFLECTORS,

BEING A DESCRIPTION OF THE

NEW ASTRONOMICAL TELESCOPES

WITH

SILVERED-GLASS SPECULA;

AND

INSTRUCTIONS FOR ADJUSTING AND USING THEM.

BY

JOHN BROWNING,

F.R.A.S., F.M.S., & F.M.S.L.

SIR WILLIAM HERSCHEL,
From an original Seal in the possession of R. W. S. LUTWIDGE, ESQ.

PRICE ONE SHILLING.

LONDON:
PRINTED BY WILLIAM FRANCIS 51, GRAY'S INN ROAD, W.C.
1867.

CONTENTS.

INTRODUCTION

Those who have only seen the heavens with the naked eye, would do well to examine various portions of the sky on a fine clear night with an ordinary opera-glass.

Many hundreds of stars will thus become visible in the milky-way; several stars that before appeared single, will be found double, and the nebulæ in Andromeda and Orion will be seen, if these constellations should be at a sufficient elevation above the horizon.

The mountains and craters on the shaded edge of the moon, in the hollow of the crescent, will also be discerned, although not well made out. As all this can be done with an instrument, magnifying only at the most, three diameters, the observer may imagine what can be seen with an instrument, that will magnify 300 diameters, (90,000 times superficial,) and such a power, the smallest telescope described in this list, is supplied with, and will perfectly bear.

The greatest number of stars visible to the unaided eye, on a clear night, is estimated at less than 3000, while the number visible by the aid of a powerful telescope, has probably been under-estimated at 10,000,000.

In one part of the heavens in the constellation of Gemini, where seven stars are visible to the naked eye, the same space is represented, on a telescopic star-map constructed by M. Chacornac, as containing 3205 stars, and these were seen with a telescope of only 6 inches aperture.

Sir William Herschel estimated the number of stars in the bright zone, known as the milky way, at eighteen millions, and Chacornac considers this estimate far too small.

It is well known that nearly the whole of Sir William Herschel's numerous and important discoveries were made with reflecting telescopes.

More recently great additions have been made to our knowledge of the celestial bodies by Lord Rosse, Lassell, Nasmyth, and De la Rue, all working with reflectors. Why, then, have reflectors fallen out of general favour, and why are they now regaining their true place in public estimation?

To see well the markings on the planets, or to observe nebulæ, perhaps the most interesting, certainly the most inexplicable of all the celestial objects, telescopes of large aperture are indispensable; an aperture of six inches being almost the smallest that can be employed successfully in viewing them. Yet the great cost of a good achromatic object glass, of even six inches dia meter, places it beyond the reach of all but wealthy persons.

For exploring the wonders of the heavens, the greater cheapness and portability of reflecting telescopes of large aperture would probably have led to their more general adoption; instead of the achromatic or refracting telescopes usually employed, but from many specula having been made of imperfect parabolic figure—from their having been fitted up without proper precautions to prevent injurious flexure—from their having been generally placed in wooden tubes, and frequently mounted on unsteady stands, and finally, from incorrect planes having been used as diagonal mirrors, their performance was unsatisfactory. The introduction of silvered glass mirrors, and of the methods of mounting subsequently described in these pages, have entirely removed these objections.

Astronomical telescopes, constructed with silvered glass specula, possess the following important advantages :—

They are only half the length of achromatics of the same aperture.

Their dividing power, on many stars, is superior to that of achromatics, aperture for aperture, as they give smaller discs ; and on the moon and planets their performance cannot be surpassed.

They are quite free from the aberration of colour.

When viewing objects in the best position for observation, that is between the zenith and 45° above the horizon, they are much more convenient to use than refractors, because the eye-piece of the reflector remains nearly horizontal ; a refractor, under the same circumstances, necessitates the observer's almost lying down on his back. Micrometric measurements may be obtained with great exactitude, from the easy position of the observer.

The price of the specula in small sizes is only about one-fifth ; and, in large sizes, only one-tenth, of that charged for good achromatic object-glasses of the same aperture.

For instruments of equal power the new reflecting telescopes do not require an observatory half the size that would be necessary for a refractor, and a building half the size would not be one quarter the expense.

These advantages, when known, must lead to the universal adoption of this instrument by that numerous class of astronomers who require powerful telescopes, but cannot afford the great outlay at which, only, refractors of large aperture are obtainable.* The Moon Committee of the British Association has recommended that those observers who desire to assist them in their labours, should observe with a power of 1000 diameters. An object glass which would bear this power with advantage, would cost from £200 to £300. A silvered glass speculum, which would bear this power, could be obtained for £35. From this it is easy to see that when silvered glass specula are more generally used, the number of observers able to employ high powers will be enormously increased.

In a paper, by Mr. Wray, recently read before the Astronomical Society, on the correction of the secondary spectum in achromatic object glasses, the writer claims that one of his new and perfected object glasses, " gives stellar images exactly similar to those shown by a well corrected reflector." This from a clever practical optician, I regard as the most favourable testimonial yet written in praise of good reflectors.

Those who have been used to observing, will understand the relative powers of the specula of various apertures, by the statement of their performance upon the double stars just enumerated ; but a better idea of their respective powers will be conveyed to those unused to working with the telescope, by stating their

* Although I am aware my motives may be misconstrued, I feel it only right to warn intending purchasers of telescopes against being misled by the numerous advertisements of cheap telescopes which are continually appearing in the public papers. The various objects they are warranted to show, are such as can be seen with any telescope having an object-glass of three inches aperture ; no objects are mentioned which would be tests of performance in definition. I have had to replace the object-glasses of several of these so-called cheap telescopes with others. The magnifying powers of the eye-pieces supplied with such instruments are often absurdly exaggerated. In one brought to me for alteration, I found an eye-piece engraved 70, only magnified 13 diameters. Of this telescope the owner has recently written to me, " I am afraid I shall not be able to do much now with your new object-glass, for the stand is as shaky as a fishing rod."

respective performance upon a planet. For this purpose I select Saturn. On this planet the 6½-inch
speculum will show 4 satellites, Ball's division in the
ring—the inner dark ring, known as the crape ring,
the shadows on the ring, and the globe, and the belts
on the globe.

The 8½ inch speculum will show 6 satellites, and in
addition to all the foregoing details, the gradations of
light on the rings, and under the most favourable
circumstances the division in the outer ring.

These particulars will be understood, by referring to
Mr. Warren de la Rue's beautiful drawing of Saturn,
made with the 13-inch reflector of his own make, which
I have inserted. (See frontispiece.)

I have also appended a drawing of Mars, as I have
seen it in Mr. Barnes' 8½-inch silvered glass reflector,
equatorially mounted, made by me.

It is not pretended that the details just mentioned
will be easily seen at any time, by an untrained observer,
but practised observers, under favourable circumstances,
will certainly be able to do more than is here stated.

To bring out the full powers of these or any other
telescopes, the objects must be favourably situated—
the air must be fine, that is, clear and steady; and
beyond this, the observer must have had some practise
in seeing minute details with low powers.

It seems but little understood that the eye is capable
of being trained no less than the hand.

I have been frequently asked to state the power of
these reflecting telescopes, as compared with achromatic telescopes. From a comparison of the results of
several observers, I have concluded that in light-grasp-

ing power, they are equal to an achromatic, one-sixth less in diameter ; while as regards their dividing power —that is, their power of separating double stars—they are fully equal to the finest achromatics of the same aperture.*

The 4½-inch, with powers from 100 to 150, will divide :—

β Orionis. α Lyræ.
δ Geminorum. ε Hydræ.
ξ Ursæ Majoris.
ε Bootis.
ν Ceti. ε Draconis.

The 6½ will divide, with powers from 200 to 300 :—

ε Arietis. α Herculis.
ξ Bootis. 32 Orionis.
ι Equulei. η Coronæ Borealis.
36 Andromedæ.

The 8½, with powers from 300 to 350, in a favorable state of the air will divide :—

γ² Andromedæ.
μ Bootis.

These last-named double stars† are both under half a second apart, and are so difficult to divide as to have hitherto, been considered good work for a 12-inch speculum.

For the instructions for silvering glass specula, contained in Appendix I., I am indebted to the kindness of Mr. With.

---

* Since writing the above, the Rev. T. W. Webb has kindly informed me that he has found a 6¼ inch silvered glass mirror, had the advantage on a light test over his 5½ achromatic, by Alvan Clark.

† I have not been able to find that these stars have ever been fairly divided by any achromatic of less than eight inches aperture, and that of admirable quality.

For an extensive list of double stars, clusters, and nebulæ, the writer can recommend the Rev. T. W. Webb's charmingly written little book " Celestial Objects for Common Telescopes," of which a second edition is now in preparation.

DESCRIPTION OF THE SILVERED GLASS REFLECTING TELESCOPES.

These telescopes are of the kind called Newtonian, a form so well known, that it is, perhaps, scarcely necessary to describe it; but I append a plain diagram (Fig. 7) and brief description, because it will assist in making clearer the instructions I have given further on, of the method of adjusting the instrument. The Newtonian telescope consists of a tube closed at the lower end, which is occupied by a concave mirror M. The cone of rays reflected from this mirror is again reflected at right angles from the surface of a small plane mirror, $m\ n$ mounted at an angle of 45° near the open end of the tube, into the eye piece, which is exactly opposite. The path of the rays is shown in the diagram (Fig. 7).*

In reflecting telescopes as originally constructed, the concave mirror was made of an extremely hard alloy, known as speculum metal. These metallic mirrors possessed several disadvantages so serious in character, that they have, for some time, fallen out of general use. The principal defects were the following :

1. From the extreme brittleness of the alloy they were very liable to fracture, sometimes breaking merely from a sudden change of temperature.

2. From their great weight it was extremely difficult to mount them in such a way as to prevent flexure, the smallest amount of which greatly injured their optical performance.

3. Their greatest drawback, however, consisted in

---

* The mirror must not be worked to a spherical, but to a very perfect parabolic curve. Those desiring information on this rather abstruse subject may read Appendix II.

the fact that the surface of the metal, from damp or other causes, sometimes became very rapidly tarnished, and this tarnish could seldom be removed, except by repolishing, and, consequently, refiguring the mirror; and this involved nearly as great an outlay as the purchase of a new speculum, besides incurring the serious risk of a fine figure being irretrievably lost.

In the telescope now described, the metallic mirror is replaced by one of glass, on the surface of which a coating of pure silver has been deposited by Liebig's process.

These glass mirrors are not at all injuriously affected by change of temperature, and their lightness very considerably reduces their liability to flexure; indeed, mounted in the manner I shall presently describe, no flexure has ever been observed in them. I may, however, state that I make the discs of the specula, which Mr. With parabolises for me, out of glass nearly twice the substance of that generally used for the purpose. The coating of pure silver reflects fully one-third more light than the best speculum metal, as the alloy before mentioned is called. But the greatest superiority of silvered glass over metallic mirrors consists in the fact that should they become tarnished, their brilliancy may readily be restored by gentle friction with soft leather, and a little of the finest rouge; and even should the silver coating become utterly spoiled, it may be easily removed, without in any way impairing either the figure or polish of the glass speculum, and a fresh one deposited at a trifling cost, thus making the mirror equal to new; and this may be repeated indefinitely. Should the owner possess a little patience, he may renew the coating himself, at the cost of only a few pence. The silvering process is fully described in an appendix.

With this alteration these telescopes have, latterly, rapidly gained ground in the opinion of practical observers, well known in the scientific world, who have had considerable experience in working with them.

On Figuring Specula.

About three years since, the Rev. Cooper Key discovered a more simple method of parabolising the surface of specula than any which had hitherto been employed, and by this process he produced two fine specula of twelve inches diameter.

The process by which these specula were worked, Mr. Key communicated to Mr. G. With, and after having worked by Mr. Key's process until a few months since, Mr. With at length contrived another plan of working, by which he considers still finer results are with greater certainty secured.

The wonderful perfection of Mr. With's specula is now generally admitted, and it is almost certain that they surpass any that have previously been produced. I have great pleasure in stating that specula of Mr. With's parabolising are now only to be obtained from me.

On Mounting Specula.

It has elsewhere been suggested that much of the dissatisfaction which has been expressed by those who have used reflectors, has arisen from their having been imperfectly mounted.

Because specula are much cheaper than achromatic object glasses, it has been supposed that they could be mounted at proportionately less cost than that incurred in mounting reflectors. This is only true to the extent that cost can be saved by reason of their shorter focal length.

It cannot be too strongly enforced that, to give the best performance, reflectors require to be mounted more steadily than refractors, because by a well-known law in optics, the effect of any vibration will be multiplied many times. Their tubes must also be carefully arranged, so as to avoid, as much as possible, the interference of air currents, which are the bane of reflectors improperly mounted or badly situated. The specula in the telescopes now described are mounted rigidly, on a new plan, which ensures permanence in adjustment, and prevents flexure. This plan is represented in Fig 1.

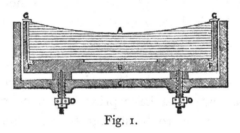

Fig. 1.

The bottom of the speculum A is a carefully prepared plane surface, and the bottom of the inner iron cell B, on which it rests, is also a plane. The speculum is clamped down in this cell by the ring G G, and it may be removed from, and replaced in, the telescope, without altering its adjustment. The elastic methods of mounting the speculum, which have hitherto been employed, generally required re-adjustment whenever the speculum had been removed. The reflecting diagonal prism, or mirror, is mounted in the manner shown in the diagrams 2 and 3.

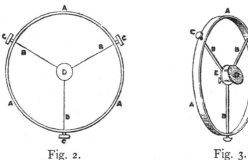

Fig. 2.   Fig. 3.

In these B B B represent strips of strong chronometer spring steel, placed edgewise towards the speculum, by which the prism or small mirror D is suspended.

The mirror thus mounted, does not produce such coarse rays on bright stars, as when it is fixed to a single stout arm ; it is also less liable to vibration, which is very injurious to distinct vision, or to flexure, which interferes with the accuracy of the adjustments.

If an observer determines to lay out a given sum in the purchase of a telescope, he will find it to his advantage to have a smaller speculum completely mounted, instead of a large speculum imperfectly mounted. With the smaller and perfect instrument he will really do more work, and with much greater comfort and satisfaction to himself. No matter how good a speculum may be, nothing can be told of its performance on difficult double stars if it is mounted on an unsteady stand.

THE SMALL ALT-AZIMUTH.

The Alt-azimuth stand, represented in Fig. 4, is entirely of iron. The tube of the telescope is of extremely stout block tin, coloured dark green, the stand being coloured dark chocolate. The body is equipoised, so that it will remain in any position, while the movements are so smooth, and the leverage so arranged, that a star may be followed, even with a power of 300, without screw motions. The instrument can be used on a table, at any window, and a stand is supplied with it, on which it can be supported at a convenient height, when it is used in the open air. This mounting is only adapted for a small-sized speculum, say not exceeding 5 inches in diameter, as, if made of a larger size, it would be so heavy as not to be portable; while with higher powers than 300, such as specula of 6 inches, and above, will easily bear, the celestial bodies cannot be followed without screw motions. By fastening the circular foot down on a block of wood of a wedge form, the angle being the complementary angle to the latitude of the place, this stand can very readily, and at a comparatively trifling expense, be made to move equatorially, so that the heavenly bodies can be followed with a single motion of the telescope. Such an arrangement is shown in Fig. 4*.

Fig. 4*.

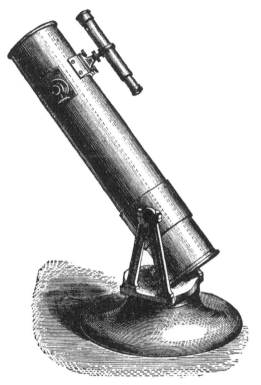

Fig. 4.

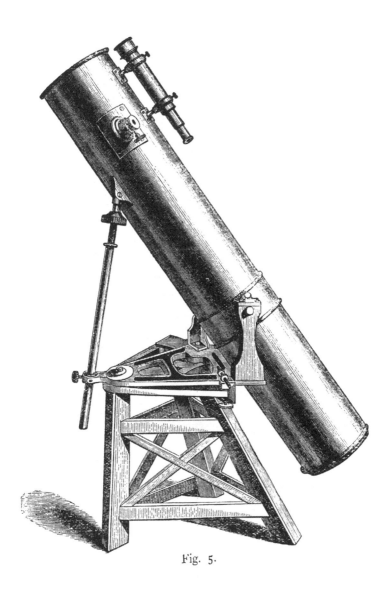

Fig. 5.

THE LARGE ALT-AZIMUTH.

The Alt-azimuth stand, of which Fig. 5 is a sketch, is well adapted for specula of 6 inches diameter, and upwards. It has quick, and fine screw motions, both in altitude and azimuth, by means of which the motions of the heavenly bodies may be followed with facility. A reference to Fig. 5 will soon make it clear how these motions are obtained. On unclamping the small screw which projects on the left hand side, on a level with the top of the stand, the telescope can be raised, or lowered, to any extent, and on reclamping the small screw it will be retained in the desired position. The fine motion in the same direction is given by simply turning round the large milled head, which is shown on the under side of the telescope, near the top. To move the telescope horizontally, a clamp screw must be released, which is attached to the tangent screw, shown on the right hand side of the stand, just above the arc of the instrument, when the telescope can be moved to any part of the arc. The fine motion, in this direction, is obtained by turning the tangent screw. This is done by means of a square key, which fits on the end of the tangent screw, the key having a long handle, and being furnished with a Hook's joint, which enables it to be turned freely when nearly at right angles to the tangent screw. The telescope is equipoised on trunnions, and it can be instantly taken off the stand at pleasure. The whole of the stand, telescope, and mountings are of metal, excepting the legs, which are very strongly made of wood, well braced together. It is exceedingly steady, indeed, as steady as any stand can be made, without adding enormously to its weight.

THE EQUATORIAL.

The stand, shown in Fig. 6, is known as an equatorial mounting. In a stand made on this plan, when the centre under the inclined hour circle, round which the telescope turns,* is directed towards the pole of the heavens, to which it will exactly point, and a star is brought into the field of view, it may afterwards be kept in view by communicating a motion in one direction to the telescope.

Again, on finding the position of any star or planet, in an almanack, and setting the two verniers on the instrument to the position given, allowing for the difference of time from the object being on the meridian, the desired object will, if the stand has been correctly adjusted, be found in the field of view. In this manner stars may be found, and well seen in the daylight, even when the sun is shining very brightly.

With Mr. Slack's 6½-inch mirror I have seen ϵ Bootis well divided, and the colours finely displayed in sunlight. On one favorable occasion I saw γ^2 Andromedæ divided with this mirror.

Those who have equatorial stands will often find about an hour before sunset a good time for observing difficult stars. Mr. Glaisher has deduced as one of the results of his scientific balloon ascents, that about this time, the atmosphere, to a great elevation, has nearly an uniform temperature.

In this Equatorial stand I have endeavoured to combine compactness and steadiness, and withal, the utmost economy with which perfect efficiency could be secured;

---

* This must not be confounded with the declination axis, which is always at right angles to it, and is shown in the diagram parallel with the top of the stand.

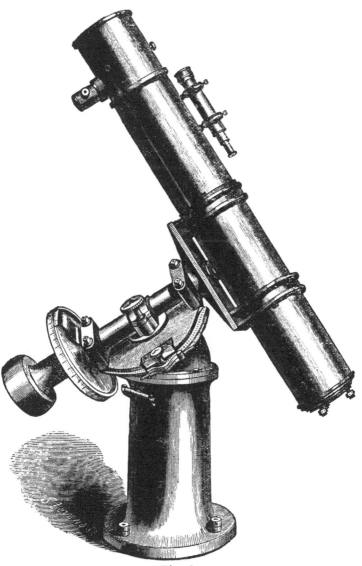

Fig. 6.

complete steadiness being the principal object aimed at in its construction.

This has been secured by making the hour circle of larger diameter than usual, and giving it a carefully ground bearing round its extreme edge; and by making the declination axis, to which the telescope is attached, much stouter than it is generally made, in proportion to the size of the telescope. The divided circles are of brass, the rest of the stand is of iron.

The upper portion of the body, containing the eye-piece and prism revolves, so that the eye-piece can always be placed in a convenient direction for observing.

While the revolving eye-piece gives great facilities for using the telescope in various positions, it is extremely difficult to adjust the prism so accurately, that in turning the eye-piece the image shall not have a slight motion in the field of view. Such a motion will alter the readings on the circles.

This difficulty may, however, be obviated, by reading the circles with the eye-piece always in one position, and by noticing any error in the places of stars, observed when it is in that position, which will be a constant quantity so long as the adjustments are not changed.

As the method of adjusting this stand is the same as in those generally constructed, it would unnecessarily occupy space if I were fully to describe it here. The reader requiring such details cannot do better than consult the valuable Hand-book of Practical Astronomy, by Mr. G. F. Chambers, taking care to obtain the new edition published by McMillan, and issued by the Clarendon Press. at Oxford.

This very handsome and admirable text book profusely illustrated contains an enormous amount of information very clearly arranged. It might well be termed the amateur Astronomer's *vade mecum.*

TO ADJUST A REFLECTING TELESCOPE, AS MADE BY JOHN BROWNING.

If the speculum has been removed from its cell, carefully dust the under part, and free it from grit, as well as the bottom of the cell, before replacing it, then screw on the ring which confines it in the cell. Before screwing on the ring, should the speculum shake in the cell sideways, insert a few long slips of stout white paper round the edge, between edge of the mirror and the cell. Having placed the mirror and cell in the telescope, and secured it by turning it in the bayonet joint, *with the cover on the mirror*, remove the glass from one of the eye-pieces, D Fig. 7, and screw it into the eye-tube.

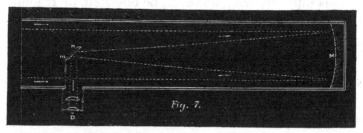

(To Adjust the Diagonal Mirror or Prism.)

Now, looking through the eye-tube, move the diagonal mirror, *m n* Fig. 7, by means of the two motions which are provided, until the reflected image of the cover of the speculum is seen in the *centre* of the small diagonal mirror, or prism.

To do this, loosen the milled-headed screw behind the

mounting of the diagonal mirror, turn the mirror until the image of the speculum cover appears central in one direction, and reclamp it by means of the screw.

The screw, close to the back of the plate on which the mirror turns, will enable the reflected image to be brought central in the other direction, and on clamping this screw the adjustment will be correct.

(To Adjust the Speculum.)

Next take the cover off the speculum, and replace the speculum in the telescope. Then move the screws at the bottom of the outside cell, which contains the mirror, until on looking into the eye-tube, the image of the small diagonal reflector is seen in the centre of the reflection of the speculum. The large hollow screws serve to move the mirror ; the small screws, which run through their centres, clamp it in position without straining it. Having clamped these screws, the adjustment will be completed, and the telescope may be turned upon an object, and focussed in the same manner as an ordinary refractor.

(To Adjust the Finder.)

To adjust the finder, get any strongly marked terrestrial object, at a great distance, into the centre of the field of view of the *large* telescope, with a tolerably high power. Then, looking through the small telescope, move it by means of the screws in the rings, until the object is exactly bisected by the cross wires in the field of view.

Now perfect the adjustment by carrying out a similar operation, using a bright star instead of a terrestrial object.

TO USE A REFLECTING TELESCOPE.

The principal difficulty in using reflecting telescopes arises from currents of air being formed in the tube, which produce unsteady vision.

These tube currents are almost entirely avoided in the telescopes now under consideration, by making the tube of iron, which quickly equalises the temperature within and without the tube. Reflecting telescopes perform best in the open air; but should an observatory be built for their reception, it should be constructed of sheet iron, as, when this material is used, the temperature within and without the building will be always nearly equal, and annoying air currents, which would be generated in a building made of non-conducting material, will be much reduced. The observing-room should be larger than will just allow of the telescope being moved about freely. When more than usually difficult work is attempted only one person should be in the observatory : by having a second person present the temperature of the interior of the building will be augmented ; also the slightest movement on the part of a second person will produce perceptible vibrations when high powers are being used.

An equatorial reflector with an aperture of $8\frac{1}{2}$-inches, I have made for Mr. R. W. S. Lutwidge, that gentleman has had placed upon the porch of his house. It is protected by a moveable cover, Fig. 7*, and is thus used in the open air. This telescope will bear a power of 500 in a state of weather when another of the same kind placed in an observatory, with a narrow slit in the dome, will scarcely bear a power of 200.

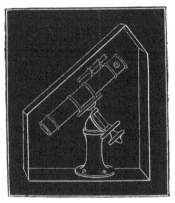

Fig. 7*.

The kind of observing room best adapted for reflecting telescopes is that proposed and adopted by Mr. Slack. This contrivance consists of a glass room of an oblong square form, each side furnished with six sliding glazed frames reaching from the floor to the roof. One half of the roof slides back under the other. There are four sashes in front of this construction, two larger and two smaller, and by different arrangements of the whole set, every part of the field of view can be uncovered in succession, and the whole chamber is easily reduced to the same temperature as the external air.

Whatever plan be adopted, at least half the roof must be open when observing. The revolving dome, with a single slide or shutter in it, is quite inadmissible for use with reflectors.

The writer has contrived a dome which consists of two segments, each rather more than half a circle; one of these segments passing over the other. In this manner an opening of any desired extent may be produced at will. To obtain a zenith view the dome should be placed rather out of centre with the building or the telescope.

Before observing, the instrument must be as much exposed to the air as possible for some time, to equalise its temperature with the surrounding air.

The temperature of the observatory when observing must never be allowed to exceed that of the outer air by more than one degree; and even if it exceeds it by only half a degree high powers cannot be used with the best results.

The use of a diaphragm one quarter less than the full aperture will improve the performance of the instrument on some objects, such as stars of the first magnitude, and also on objects generally when the air is exceptionally unsteady.

The performance of a telescope should never be condemned from a single trial, or, indeed, after many trials. From unsteadiness of the air, the finest instruments that will at times divide stars less than half a second apart, will not, upon some nights, separate stars two or three seconds asunder.

The air is, generally, the most unsteady from inequality of temperature about an hour after sunset. This time is, therefore, ill-adapted for observing particularly difficult objects. In London, the air improves greatly in steadiness as well as clearness, after midnight. After rain has fallen the air is generally, both clear and steady, and therefore well adapted for observing.

ON PRESERVING THE SILVER SURFACE OF THE SPECULUM.

Several methods, depending, on the employment of deliquescent chemical substances, have been proposed for preventing the silver surface from becoming tarnished. These methods are troublesome, and sometimes do more harm than good. Practically, I find it quite sufficient to keep the speculum covered with a tightly fitting cover, when not in use ; and with such a cover all mounted specula should be provided.

It is a good plan to envelope the telescope, when not in use, in a cover made of American leather cloth. Should the speculum be left in its place, this covering serves to protect the silver film, and, under any circumstances, it will serve to keep dust out of the instrument. If in spite of precautions, a deposit of moisture should take place on the speculum, or rain be allowed to fall on its surface, the mirror should be placed in front of the fire, taking care not to make it too hot, and kept there until the moisture has evaporated, and the surface become perfectly dry. If any stains should be left, they may be removed by polishing in small circular strokes, as in Fig. 11, with a rouged leather pad, first letting the mirror cool. The pad should be warmed and let to cool before using.

No matter how bad the surface of the mirror may appear if these instructions be carefully followed, it may be restored to its original brilliant condition, but *if the moistened surface be rubbed before it has become perfectly dry, the whole of the silver surface will be removed.*

The speculum should not be taken from the cold air to a warm room, or when this cannot be avoided, the cover should be placed on the speculum while it is in the open air, or cold place, and the speculum and cell put in a box, with a well fitting cover, before it is removed to the warm apartment. This will prevent a deposition of moisture.

When carefully used the silver surface should last without renewal for three or four years.

APPENDIX I.

TO SILVER GLASS SPECULA.

Prepare three standard solutions :—

Solution **A** { Crystals of Nitrate of Silver, 90 grains. } Dissolve.
{ Distilled Water. . . . . 4 ounces. }

Solution **B** { Potassa, *pure by Alcohol* . . 1 ounce. } Dissolve.
{ Distilled Water . . . . . 25 ounces. }

Solution **C** { Milk-Sugar (in powder) . . ½ ounce. } Dissolve.
{ Distilled Water . . . . . 5 ounces. }

Solutions **A** and **B** will keep, in stoppered bottles, for any length of time, solution **C** must be fresh.

THE SILVERING FLUID.

To prepare sufficient for silvering an 8-inch speculum :—

Pour 2 ounces of Solution **A** into a glass vessel capable of holding 35 fluid ounces. Add, drop by drop, stirring all the time (with a glass rod), as much liquid ammonia as is *just* necessary to obtain a clear solution of the grey precipitate first thrown down. Add 4 ounces of solution **B**. The brown-black precipitate formed must be *just* re-dissolved by the addition of more ammonia, as before. Add distilled water untill the bulk reaches 15 ounces, and add, drop by drop, some of solution **A**, until a grey precipitate, which does not re-dissolve after stirring for three minutes, is obtained, then add 15 ounces more of distilled water. Set this Solution aside to settle. Do not filter.

When all is ready for immersing the mirror, add to the silvering Solution 2 ounces of Solution **C**, and stir gently and thoroughly. Solution **C** may be filtered.

Perfectly pure chemicals may be obtained of Messrs. Jackson & Townson, 89, Bishopsgate Within, London, E.C.

TO PREPARE THE SPECULUM.

Procure a circular block of wood 2 inches thick and 2 inches less in diameter than the speculum. Into this should be screwed three eye-pins, at equal distances, thus :—(Fig. 8). To these pins fasten stout whipcord, making a secure loop at the top.

Melt some soft pitch in any convenient vessel, and having placed the wooden block face upwards on a level table, pour on it the fluid pitch, and on the pitch place the back of the speculum, having previously moistened it with a thin film of spirit of turpentine to secure adhesion. Let the whole rest until the pitch is cold.

TO CLEAN THE SPECULUM.

Place the speculum, cemented to the circular block face upwards, on a level table, pour on it a small quantity of strong nitric acid, and rub it gently all over the surface with a brush made by plugging a glass tube with pure cotton wool (Fig 9). Having perfectly cleaned the surface and sides, wash well with common water, and finally with distilled water. Place the speculum face downwards in a dish containing a little rectified spirit of wine* until the silvering fluid is ready.

TO IMMERSE THE SPECULUM.

Take a circular dish about 3 inches deep, and 2 inches larger in diameter than the speculum. Mix in it the silvering solution and the solution **C**, and suspend the speculum, face downwards. in the liquid, which may rise about $\frac{1}{4}$ of an inch up the side of the speculum.

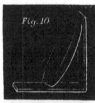

When the silvering is completed† remove the speculum from the solution, and immediately wash with plenty of water, using at least 2 gallons, and finally with a little distilled water. Place the speculum on its edge on blotting paper to drain and dry (Fig 10).

When perfectly dry, polish the film by gently rubbing first with a piece of the softest wash-leather, using circular strokes (Fig 11), and finally with the addition of a little finest rouge.

A "flat" may be silvered by fastening with pitch to a slice of cork, cleaning as above described and using as much silvering fluid as will form a stratum about $\frac{1}{2}$ inch deep beneath the mirror.

---

* The silvering will be completed in from 50 to 70 minutes, according to temperature; 50 minutes will be sufficient in summer.
† Not methylated.

TO SEPARATE THE SPECULUM FROM THE BLOCK.

Stand the speculum on its side, insert the edge of a sharp half-inch chisel between the wood and glass, administering two or three gentle blows, and the block and mirror will separate safely and easily. It is preferable to obtain the aid of an assistant in this operation. Any pitch which remains on the back of the mirror may be removed by scraping, and a little turpentine.

The cost of silvering an 8-inch speculum, exclusive of the cost of alcohol, which may be used over and over again, will not exceed 9d.

Nitrate of Silver being 4s. per oz.

Potash . . . . . 8d. „

Milk, Sugar . . . 2d. „

Avoid all excess of ammonia, and be sure to use *pure* distilled water.

APPENDIX II.

"ON WORKING GLASS SPECULA."

WHEN parallel rays of light are allowed to fall upon the surface of a concave mirror, if the surface be a sperical curve, the rays will not all be reflected to a single point.

In (Fig. 12) it will be seen that the rays A, falling on the mirror, would be reflected and form an image at *a* ; while the rays B B would be reflected and form an image at *b*, farther from the front of the mirror.

If the reflected images were viewed with an eye-piece placed any where in front of the mirror, they would not be in focus at the same time, so that only a blurred and indistinct image would be seen.

To make   e mirror reflect rays falling on all parts of its surface to one point, it is necessary that it should be fashioned into a parabolic curve.

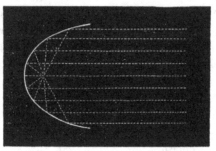

Fig. 15.

Such a curve is shewn in Fig. 13, which may be considered as a spherical curve, in which the curve has been made deeper, or the outer portion flattened. In practice, the amount of this difference is so exceedingly minute as to be inappreciable by actual measurement.

Sir John Herschel states that the utmost variation of a four-foot speculum from a spherical curve is less than one 21,000th part of an inch. Yet it is well known that for telescopic use a mirror with a spherical curve is, for the reason just explained, totally useless.

In working the glass specula, a disc of hard crown glass, varying in substance from three-quarters of an inch to one and a-half inches, according to the size of the speculum for which it is intended, is turned, and polished on the edge. One side of this disc is ground to a truly plane surface. On this side the speculum, when mounted on the writer's plan, rests in its cell. The other side is then ground to a concave spherical curve of such a radius as will produce the desired focus. This spherical curve is converted into a parabolic figure somewhat thus:—

An iron tool, similar to that on which the spherical curve has been ground, is covered with a layer of pitch, tempered to a certain consistency. This pitch is warmed, and the speculum being laid upon it, makes the pitch assume the same curve. The speculum is then polished on the pitch with rouge. In this polishing, the speculum and polisher are not worked together equally all over the surfaces, but the speculum is moved in such a manner that it is polished away most towards the edge, and a parabolic curve is produced. During the process, both the speculum and the polisher continually revolve.

The Diagram of Lord Rosse's machine, with which he figured his speculum six feet in diameter, will give an idea of the action of the speculum and polisher on each other.*

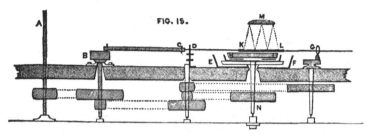

FIG. 15.

This machine is represented in Fig. 15; A is the spindle, by turning which the whole machine is set in motion; H I is the speculum; K L the polisher; B an excentric which carries the polisher backwards and forwards; G another excentric which moves the polisher from side to side slowly, during the reciprocating motion. The amount of motion given to the polisher, and the rapidity of rotation of the speculum can be changed at pleasure.

FIG. 14.

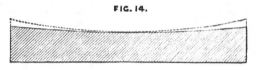

In Fig. 14 the dotted line represents the spherical curve of the mirror when the polishing is begun, and the continuous line the parabolic curve it assumes when the polishing process is finished. It will be of course understood that, in all the diagrams, these curves are enormously exaggerated.

During the graduated polishing, the speculum is repeatedly tested, until the desired definition is attained. When completed, if accurately figured, the marginal inch of the speculum should give equally sharp definition with the centre, and have identically the same focus.

In figuring the mirrors of the telescopes herein described, an improved method has been adopted of fashioning the parabolic curve; it is believed this method gives superior results to any hitherto attained.†

* This diagram is copied from Sir John Herschel's work on "The Telescope."
† The reader who wishes for further information on this subject is referred to Sir John Hershel's work on "The Telescope."

NOTICES WHICH HAVE APPEARED OF THE NEW TELESCOPES.

In the means of suspending the small mirror, the suppression of a great part of the thickness of the arm tends to do away with the rays which we all know appear to shoot out from the image of a star in the reflector.— WARREN DE LA RUE, ESQ., F.R.S., As President of the Royal Astronomical Society.

MR. BROWNING exhibited a silvered-glass speculum equatorially mounted in a very handy form.—ASTRONOMICAL REGISTER.

These two contrivances of Mr. Browning—the methods of mounting the speculum, and the support for the diagonal plane mirror or prism—appear from the trials given to them in Mr. Slack's telescope to answer their purpose exceedingly well.

The President of the Astronomical Society expressed a doubt whether the cellular plan of mounting the mirror would answer for large instruments, but if it performs well up to seven, or eight, or ten and a quarter inches, the last being the largest size to which Mr. Browning has yet adapted it, its importance will stand very high. The new system of mounting the prism or flat has great advantages. The three slender springs do much less optical mischief than the single stout arm previously employed, and contribute with the excellent working of Mr. With's mirrors to remove the defective definition which reflectors usually give of large stars. In Mr. Slack's instrument the definition closely resembles that of a fine refractor, and the discs are remarkably small. In this equatorial stand, the special ends in view were great stability with convenience and economy.

Hitherto moderate priced stands have usually been of comparatively slight construction, and though many of them possess considerable merit, none of them seem adapted to carry a somewhat heavy and bulky telescope. It will be seen from the drawing that the base of the new stand is very compact and solid; it is, in fact, a stout cast-iron tube. The circles are twelve inches in diameter, reading to 1 minute of an arc, and 2½ minutes of time. The declination circle has considerable weight, and thus effectively assists in counterpoising the telescope. The tube of the telescope divides into two parts, each furnished with a flange fastened by screws to stout rings, supported by a heavy arm ; by this means the principle weights are exactly opposite each other in every position of the instrument, and they are kept near the centre of the polar axis, and at about equal distances from the centre of gravity of the pillar stand. The hour angle motion has bearings equal to the diameter of the hour circle, twelve inches, which adds to the stability. Upon trial, this telescope is found to be remarkable steady and free from vibration under a power of between 600 and 700, and the result of this steadiness is very conspicuous in the definiteness of the division of double stars, when the lowest powers are employed that can produce such a result.

The eye-piece and prism, or flat, revolve so that the awkward positions to which an observer is subjected when an ordinary reflector is mounted equatorially, are completely obviated. That these silvered glass telescopes will come into favour, cannot be doubted, as they cost only a fraction of the price of refractors, capable of doing the same work, and perform to the satisfaction of performers like Mr. Webb—who has tried a good many—Mr. Cooper Key, Mr. Bird, and others.—INTELLECTUAL OBSERVER.

LIST OF PRICES.

SILVERED GLASS SPECULA UNMOUNTED.

The performance of these specula will be guaranteed; they will bear a power of 100 to the inch on suitable objects, and under favourable conditions of the atmosphere.

					£	s.	d.
Speculum, 4½ inch diameter, about 3 feet focus				......	4	10	0
„ 6½	„	„	5	„	8	10	0
„ 8½	„	„	6	„	16	0	0
„ 9¼	„	„	6	„	21	0	0
„ 10¼	„	„	7	„	35	0	0
„ 12	„	„	8	„	50	0	0
„ 13	„	„		„	75	0	0

PRICES OF SILVERED GLASS SPECULA ASTRONOMICAL TELESCOPES, ON ALT-AZIMUTH STANDS.

	£	s.	d.
4½ inch speculum, 3 feet focus, on iron alt-azimuth stand, with two eye-pieces, 100 and 300 (Fig. 4)	16	0	0
6½ inch speculum, 5 feet focus, on alt-azimuth stand, with quick and slow fine screw motions, and three eye-pieces, 100 to 450 (Fig. 5)	30	0	0
8½ inch speculum, 5 feet 6 inches focus, mounted as above, with three eye-pieces, 100 to 500 (Fig. 5)	40	0	0
9¼ inch speculum, 5 feet 9 inches focus, as above, with four eye-pieces, 100 to 600 (Fig. 5)..	50	0	0
10¼ inch speculum, 6 feet 9 inches focus, ditto (Fig. 5)...	70	0	0

These telescopes have plain mirrors which can be replaced with prisms by payment of the difference of the prices stated in the lists which follow.

SILVERED GLASS SPECULA ASTRONOMICAL TELE-SCOPES EQUATORIALLY MOUNTED.

	£	s.	d.
4½ inch speculum, 3 feet focus, equatorially mounted (angle for latitude to order) with 6 inch hour circle reading to 5 seconds, and declination circle reading to 1 minute, two eye pieces 100 and 300 (Fig. 6)	30	0	0
6½ inch speculum, 3 feet focus, with 12 inch hour circle reading to 5 seconds, and declination circle to 1 minute, three eye-pieces 100 to 450	75	0	0
8½ inch speculum, 5 feet 6 inches focus, mounted as above	95	0	0
9¼ Ditto 5 feet 9 inches focus, with four eye-pieces 100 to 600	120	0	0
10¼ Ditto 6 feet 9 inches focus	150	0	0

These instruments are all furnished with reflecting prisms of the finest quality, in place of the diagonal mirrors generally used.

Clock work driving apparatus to order.

SILVERING GLASS SPECULA.

	£	s.	d.
4½ inches	0	5	0
6½ ,, ...	0	10	0
8½ ,, ...	0	12	6
9¼ ,, ...	0	15	0
10¼ ,, ...	1	0	0

Diagonal Planes, 2s 6d.

All charges incurred for carriage will be extra.

ASTRONOMICAL EYE-PIECES—HUYGHENIAN CONSTRUCTION.

Nos.	1 and 2,	Magnifying 65 or 85	£0	15	0
„	3, 4, and 5,	„ 125, 200, or 250	1	0	0
„	6,	„ 400........................	1	5	0
„	7,	„ 600.............................	1	10	0

ACHROMATIC EYE PIECES.

These eye-pieces have a rather limited field, but their performance with reflecting telescopes, particularly on planets, is very superior to Huyghenian. They were first used by the Rev. T. W. Webb.

A,	Magnifying	95	£1	0	0
B,	„	160	1	5	0
C,	„	200	1	10	0
D,	„	275	1	10	0
E,	„	525	1	15	0
F,	„	850	2	5	0

All these powers are calculated on an object-glass of 5 feet focus.

DIVIDED LENS DYNAMETER.

For measuring accurately the power of eye-pieces, new movement for separating the lenses, £4 4s.

BARLOW'S LENS.

This is an achromatic combination of a negative focus; on inserting it behind any eye-piece (that is between the eye-piece and the object glass) the power of the eye-piece is increased from one-third to one-half, at the same time the introduction of this lens, when using Huyghenian eye-pieces, greatly improves the performance of Reflecting Telescopes, especially on bright stars.

Price of the best quality £1.
Second quality ....10s.

SMOKE-COLOURED GLASS WEDGES.

Corrected for refraction, for intensifying the markings on the moon or planets £1.

MICROMETER.

Micrometer for measuring to fractions of a second, £5 5s.

SILVERED GLASS PLANES FOR USE AS DIAGONAL MIRROR IN REFLECTING TELESCOPES.

1 inch to 1½ in the minor axis of the ellipse............£0 10   0

1 ½ to 2 inch   „   „   1   0   0

PERFECT PLANES.

UNMOUNTED.

PERFORMANCE UNDER ANY POWER GUARANTEED.

1   inch in the minor axis of the ellipse.................£1   0   0

1½   „   „ 1 10   0

2   „   „   2   0   0

2½   „   „ 2 10   0

DIAGONAL PRISMS.

UNMOUNTED.

With circular surfaces on the planes at right angles to each other. The diameter of these planes is equal to the minor axis of an ellipse which would be required if a mirror were employed.

			£	s.	d.
Plane surfaces ¾ inch in diameter			1	1	0
Ditto	1	„	2	2	0
Ditto	1¼	„	3	3	0
Ditto	1½	„	5	5	0
Ditto	2	„	7	7	0
Ditto	2½	„	10	10	0

These prisms are made of very pure hard white crown glass. They reflect more light and freer from color than silvered diagonal mirrors They are not liable to injury from moisture.

Prepared pad for polishing specula, in bottle, 2s. 6d.

SPECTROSCOPES.

	£	s.	d.
Amateurs' Spectroscope	2	2	0
Student's ditto	5	5	0
Model Spectroscope, with two prisms	10	10	0
ditto ditto three prisms	15	0	0
Large ditto, for physical researches, with four prisms, on			
the plan of the Gassiot Spectroscope..................	30	0	0
Ditto, with seven prisms.....................................	45	0	0
Herschel-Browning direct vision spectroscope for geologists,			
tourists, &c.	5	5	0
Ditto, smaller size, for the pocket	4	4	0

Star Spectroscopes, £7 7s.; £12 12s.; £18 18s.

Spectroscopes for microscopes, £3 3s.; £4 4s.; £5 5s.

———————

An illustrated Catalogue of Spectroscopes sent post free for
seven stamps.

———————

JOHN BROWNING,

OPTICAL AND PHYSICAL INSTRUMENT MAKER

To Her Majesty's Government, the Observatories of Greenwich, Kew
Cambridge, &c., &c.

111, MINORIES, CITY, E.   FACTORY, 6, VINE STREET, E.C., LONDON

———————

PRIZE MEDAL 1862.

———————

ESTABLISHED 100 YEARS.

———————

MICROSCOPES, TELESCOPES, OPERA GLASSES, SPECTACLES, ETC., ETC.

Printed in the United States
By Bookmasters